Quantum Magic

by

Monk E. Mind

Other Books by Monk E. Mind
www.monkemind.com

Copyright © 2018 Monk E. Mind

Grinning Monkey Publishing

All rights reserved.

ISBN-13: 978-1729681602
ISBN-10: 1729681603

Acknowledgments

Many thanks to friends of the Rational Scientific Method (RSM) facebook group from whose dedication to rational science I have drawn great inspiration.

Thanks to Bill Gaede, his book, WGDE, and for many hours of discussion between myself and the Rational Scientists.

I'd like also to thank our many detractors and naysayers for forcing me to be more precise and for helping me to realize the strength of the RSM arguments.

As always, thank you Mom and Dad, for without you I would not have been possible.

Contents

Forward

Chapter One – Gravitational and Electromagnetic Forces
Chapter Two – The Nuclear Forces
Chapter Three – What is Force Anyways?
Chapter Four – Quantum Fluctuation
Chapter Five – The Weak Nuclear Force
Chapter Six – Is Gravity a Force?
Chapter Seven – What is Quantum Gravity?
Chapter Eight – Evidence for Quantum Gravity
Chapter Nine – Quantum Experiments
Chapter Ten – Competing Quantum Theories
Chapter Eleven – Loop Quantum Gravity
Chapter Twelve – Crossing the Quantum Divide
Chapter Thirteen – Theoretical Mathematics
Chapter Fourteen – Karl Popper
Chapter Fifteen – Word Magic
Chapter Sixteen – Black Holes
Chapter Seventeen – Gravitational Waves
Chapter Eighteen – Mass and Energy
Chapter Nineteen – Time
Chapter Twenty – The War Over Reality
Chapter Twenty One – The Neutrino Puzzle
Chapter Twenty Two – Quantum Computers
Chapter Twenty Three – Math is Descriptive
Chapter Twenty Four – Rope Hypothesis
Chapter Twenty Five – Forces of Nature – Push and Pull
Chapter Twenty Six – Gravitation and Electrostatics
Chapter Twenty Seven – Closed Circuit
Chapter Twenty Eight – Photoelectric Effect
Chapter Twenty Nine – Antenna Theory
Chapter Thirty – Light, Gravity and Magnetic Moment

Monk E. Mind

Forward

It is assumed that the reader is, to some degree, acquainted with physics in general, or Quantum Mechanics in particular. It is also hoped that the reader has familiarized themselves with the Rational Scientific Method.

If you are not familiar with Quantum, I ask your forbearance as I briefly touch on different aspects of Quantum while not defining every term or explaining every concept. We are, after all, discussing at least a hundred years of particle physics.

When you read something like this: "According to Maxwell's equations, EM fields travel at finite velocity." I am using the familiar language of mathematical theorists. Rational science would never use this kind of irrational language, and that will eventually be made very clear to the reader. Field is a location, and locations, like all concepts, do not travel. Rational scientists never move concepts, we move objects. A magnetic field, for example, might be sweeping magnetic threads.

Throughout the book, I refer to other resources, such as my books Rational Science Vols. I-6 and Rope Hypothesis and Thread Theory; all available on Amazon.com. Articles, videos and books by Bill Gaede, the originator of Rational Science and Rope Hypothesis, can be found at HubPages, YouTube, Patreon, at Rational Scientific Method facebook group and his website, wgde.com. To speak directly with myself or Bill Gaede and the Rational Scientists, come by The Rational Scientific Method facebook Group.

If you are familiar with Quantum, I invite you to the Rational Scientific Method facebook group to discuss what is presented here. If you feel that anything has been misunderstood or misrepresented, please make yourself available for our edification. We welcome your discourse and your dissent. If, on the other hand, this small work peaks your interest, we also welcome you to question, debate and discuss with the luxury of detail, anything within these pages.

To everyone, I understand that there is no settled science and as such this is presented only as a beginning to what should prove to be a long discussion on what is science and what are the physical mediators of all phenomena. Welcome to the world of Rational Science.

Chapter One - Gravitational and Electromagnetic Forces

Modern physics studies the four "known forces" of nature; Gravity, Electromagnetic, Weak and Strong nuclear forces. However, "known forces" is a misnomer because hundreds of years of investigation have not lead to any clearer understanding. In short, scientists do NOT know what these forces are. In this book we will cover, with the luxury of detail, the history and current investigative nature of particle physics and the so-called forces.

What are the forces that keep matter together? What holds atoms together, the cells to our bodies, the moon to the earth, and the planets to the solar system or the stars to the Milky Way galaxy?

What does it mean when we use force to push something over, or convert energy in combustion engines to apply force to the drive train and wheels of our automobiles? What is the difference between forces that act when surfaces of objects touch or collide and forces that act at a distance like when a magnet attracts a piece of iron?

Scientists have developed basically two approaches to answer these questions: They study the components of sub atomic particles which comprise matter, and they study the forces of push and pull between them. Push is known as the electromagnetic interaction and the weak force, and pull is known as gravity and the strong force.

Gravity

Physics has always been pretty good at describing these phenomena, but, to this day, scientists are still unable to imagine a mediator for pull. Newton, in his Principia, said that gravity was the force that held the planets to the sun and described this action with his formula $F=Gm_1m_2/r^2$, but he never understood the underlying physical mechanism that mediates this attractive force. He said this:

"It is inconceivable that inanimate Matter should, without the Mediation of something else, which is not material, operate upon, and affect other matter without mutual Contact...That Gravity should be innate, inherent and essential to Matter, so that one body may act upon another at a distance thro' a Vacuum, without the Mediation of any thing else, by and through which their Action and Force may be conveyed from one to another, is to me so great an Absurdity that I believe no Man who has in philosophical Matters a competent Faculty of thinking can ever fall into it."

Later, scientists came up with the concept of a gravity field throughout space and celestial mechanics is still having a field day! In the 18th and early 19th century scientists predicted a planet outside of Uranus by irregularities in its orbit and consequently astronomers were able to discover Neptune as a result. Newton had placed a value of zero on the precession of planets in their orbit around the sun but Leverrier calculated Mercury's at a value of 35" per century. A more accurate value was later determined to be 43", but it was not until 1915 when Einstein "explained" the inconsistency.

Newton's formula for force, F=ma, described an apple (mass) falling to the ground at 9.8 meters per second squared, but didn't explain why objects fall at the same rate (or why they fall down instead of up). In Newton's gravitational formula m refers to gravitational mass (static) and in his force formula m refers to inertial mass (dynamic).

When you don't know which is gravitational mass and which is inertial mass in any particular phenomenon, you can't know which force is responsible; gravitation or acceleration. When you gun the engine of your car, acceleration is mostly responsible for pushing you back against the seat. Same for an astronaut being launched into space, but if you can't tell what is due to gravity and what is due to acceleration you can exchange gravitational force by being in an "accelerating frame." When acceleration cancels gravitational force you have an "inertial frame." This frame varies from point to point because the "gravity field" is different at each point. Einstein's Equivalence Principle states that you can find an inertial frame anywhere in space and time where gravitation is removed from the equation. Therefore, gravity is used as a coordinate system and so is considered a "geometric theory."

This is different than the flat Euclidean geometry, since it covers time and space, and so is known as Non-Euclidean Geometry. Hence, force is a result of the properties of space and time. Gravity, then, is a result of curved space time, so that an object weighs down the space time fabric and then accelerates to the lowest point taking the shortest route (geodesic) to get there.

Einstein's theory of gravitation, known as General Relativity describes a gravity field as a geometric quantity defining "proper-time" as a value used in all coordinate systems as though time were distance. Just plot time along with length, width and height and you have space time. Using Einstein's equations, Albert was able to compute the value for Mercury's precessional orbit.

Electromagnetism

The person accredited with unifying electricity and magnetism is James Clark Maxwell. The mediator of electromagnetic force is the electromagnetic field. We can divide this into the electric and the magnetic fields. These fields are not symmetric as electric fields have electric charges, but magnetic fields lack magnetic charges which are only predicted by Quantum theory to exist in extremely small quantities. A magnetic charge has never been discovered. An electric field is caused by charged particles, and a magnetic field is caused by MOVING charged particles.

A charge is a phenomenon that can produce a force field around it that can interact with other fields. Electric charge is responsible for electric fields and also magnetic fields when it is in motion.

[NOTE: We may be told that mass charge is responsible for gravitational fields and a Strong charge is responsible for strong and weak nuclear fields.]

As we learned from Rational Science Vol. V; Chapter Forty – Gravitation and Electrostatics:

Both gravity and electrostatic "force" are stronger or weaker based on "the inverse of the distance squared." The two have similar formulas:

$$F_g = G \cdot M_1 \cdot M_2 / D^2$$
$$F_e = K \cdot Q_1 \cdot Q_2 / D^2$$

We see that Newton's formula is similar to Coulomb's Law of electrostatics. The gravitational force is different in that it is only attractive whereas the electrostatic force is attractive and repulsive.

Charges are said to be either negative or positive, and, by convention, marked with a plus or minus sign to denote the difference. Positive and negative charges attract and bind together in atoms reducing the electromagnetic field around a given object so that there might be a very minute electromagnetic interaction (EMI) between a rock and a planet on which it resides. A gravity field, which is always attractive, means that every particle in the rock is interacting with every particle in the planet and creates a force equal to the weight of the rock. Compare the electromagnetic interaction (EMI) between two electrons to the gravitational force and find that the EMI is 10^{42} (about 40 magnitudes) larger than the gravitational one at this micro scale of elementary particles.

Because of this extremely large difference between the two forces, when studying physics at the sub-atomic level of elementary particles, like electrons, protons and neutrons, we can ignore the force of gravity. The study of quantum electrodynamics does this for "ordinary energy levels."

According to Maxwell's equations, EM fields travel at finite velocity. Coulomb's law is a static law and holds true only when the electromagnetic field has moved between two charges. Remer discovered that the speed of light (c) is finite and Huygens discovered that EM fields also travel as waves just like visible light. Max Planck quantized light to explain black body radiation and Hertz observed, and later Einstein got the Nobel for, the photoelectric effect (by proposing the particle wave duality of light).

Finally, Arthur Compton "proved" photons could dislodge electrons experimentally, and, from that time on, most scientists accepted that the EM field is comprised of a stream of photons that behave as a wave. Photons are point particles and therefore an electric field is a region or mathematical grid where different points represent different charges.

Looking at electric force between charges being mediated by the electromagnetic field over distance helped some to "make sense" of this mysterious invisible concept "force." Scientists could now imagine photons streaming across space and impacting other particles. This is a phenomenon we understand. If you hit someone they feel the force.

But it's really a bit more complicated than that, we are told. According to quantum, all the particles are identical. When two identical, charged particles interact, which one is sending and which one is receiving the photon? Quantum mechanics says both. The two point-like particles hit each other and the result is force.

Thank goodness for quantum mechanics! With Heisenberg's uncertainty principle, we understand that we can not measure both energy and time, or position and momentum concurrently.

Energy can not be determined because of the uncertainty in determining time, and we can not know both the position and momentum of an electron. We can only know one or the other. Therefore, we can say that photons travel from one charge to another along the force field between them because we can not know the time due to uncertainty. We can also say that a photon is massless because the relationship between Energy^2=momentum^2c^2 (energy momentum formula) need not be satisfied.

Translate energy and 3D momentum into relativity's four-momentum and we end up with virtual photons with virtual mass. Then, we can assume either a photon travels from particle A to B with a four-momentum or from B to A with the opposite four-momentum. Since Coulomb's law is accurate over long distances we assume that photons must be close to massless. In order to measure c, photons have to interact with each other. Since there is uncertainty in a photon's mass and its speed, but we measure c as constant, we can assume that the photon's mass is basically zero.

Feynman and others described the EM force between elementary particles and called that quantum electrodynamics (QED). It is a blending of special relativity and quantum physics. By the use of Feynman diagrams we return to Coulomb's law where elementary particles, such as electrons, exchange photons. The diagrams are based on mathematical expressions from which basic rules are derived for propagation of virtual particles and their interaction with vertices. A vertex is where a line intersects with another indicating where one particle emits or absorbs another.

QED: "In particle physics, quantum electrodynamics (QED) is the relativistic quantum field theory of electrodynamics. In essence, it describes how light and matter interact, and is the first theory where full agreement between quantum mechanics and special relativity is achieved. QED mathematically describes all phenomena." - wikipedia

Chapter Two - The Nuclear Forces

At one time it was believed that there were only two forces, gravitation and electromagnetism. It was supposed that electromagnetism was responsible for the forces which held the atom's nucleus together. Some thought that the nuclei were protons and electrons and also that electrons somehow kept the protons together. This was problematic because scientists wanted to know how electrons in the nucleus and electrons that orbited the nucleus were different. What bearing did the fact that electrons were compressed into the small nucleus have on Heisenberg's uncertainty principle? This principle forbids the electron from being found in the nucleus. If it was there in the nucleus, its position would be known and the margin of error would be very, very tiny. For that to be the case any measurement of its momentum would be very, very large and the electron would be moving so fast that it would fly out of the atom.

There were no other known elementary particles, but some radioactive decays revealed electrons emanating from the nucleus. James Chadwick discovered a type of neutral radiation emanating from the atom's nuclei and his experiment showed there were electrically neutral particles which were called neutrons.

Later, Wigner proposed that there were two types of nuclear forces involved: the Strong nuclear force which held the protons and neutrons together in the nucleus and the Weak nuclear force which caused radioactive decay.

Strong Nuclear Force or Interaction

Hideki Yukawa proposed a field theory with the mediating particle being the Meson with a small mass. The Electromagnetic force has a longer range compared to the Strong force which is only attractive within the radius of the nucleus. Another physicist discovered a particle from cosmic rays later called the muon, a heavy relative of the electron that reacts very little with matter.

Cecil Powell measured the pion's properties and when the Feynman diagrams were applied to it the coupling constant was larger than one. As you created more interactions (exchange of a pion), the perturbation expansion indicated the scattering of two protons which produced different strongly interacting particles called hadrons. More and more particles were discovered until Murray Gell-Mann proposed these strongly interacting particles were bound states of more fundamental particles called quarks. If you want to understand what is happening inside the nucleus you need to understand the field theory for quarks.

Weak Nuclear Force or Interaction

After the discovery of radioactive uranium salts, came the neutron. Atoms decay by emitting radiation and this is an example of yet another force. The neutron decays into a proton and an electron and, then later, a new particle proposed by Wolfgang Pauli, the neutrino. Well, it was called that, but is considered to be the anti-neutrino now. A proton can also decay into a neutron, a positron, and a neutrino since the mass of nucleons are virtual. Enrico Fermi's model suggested that these interactions were instantaneous and later Feynman and others modified the theory to account for parity violations. These weak interactions can distinguish between left and right (called parity).

The model was not renormalisable (a math trick we'll learn about later) so could not be used as a general theory but still worked well for some processes. To fix this seemingly irreconcilable problem, extremely heavy mediating particles were proposed by Steve Weinberg and others with Non-Abelian Guage Theories. These field theories were generalizations of QED and used various self interacting, mediating particles. The theories were eventually proven to be renormalisable and so labeled good quantum theories. These particles, around 100 times heavier than a proton, were discovered at CERN by Rubbia and Vander Meer.

SLAC experiments verified the existence of quarks and something called scaling proposed by Bjorken and explained by Richard Feynman. Scaling has to do with scattering of electrons and protons with fewer variables at higher energies. Protons consist of point like components whose coupling strength decreases as energy increases. This is called asymptotic freedom and is opposite QED. Quantum field theory supposed that the coupling constant was a result of the screening of virtual particle pairs. Relativistic QM allows for this pairing if they are short lived since energy is the same as mass ($E=mc^2$). We can also invoke Heisenberg's uncertainty principle again.

Quantum field theory makes it hard to believe in asymptotic freedom because it must mean that the quark charges are antiscreened. David Gross and others claim that the requirement of asymptotic freedom is satisfied with Non-Abelian Guage Theories if there are not too many quarks. Now, don't we sound smart?!

Screening has to do with how a field acts on virtual particles carrying a particular charge. According to wikipedia, referring to variations in the coupling constant at different scales, in QED:

"In the vicinity of a charge, the vacuum becomes polarized: virtual particles of opposing charge are attracted to the charge, and virtual particles of like charge are repelled. The net effect is to partially cancel out the field at any finite distance. Getting closer and closer to the central charge, one sees less and less of the effect of the vacuum, and the effective charge increases."

"In QCD the same thing happens with virtual quark-antiquark pairs; they tend to screen the color charge. However, QCD has an additional wrinkle: its force-carrying particles, the gluons, themselves carry color charge, and in a different manner. Each gluon carries both a color charge and an anti-color magnetic moment. The net effect of polarization of virtual gluons in the vacuum is not to screen the field, but to augment it and change its color. This is sometimes called antiscreening. Getting closer to a quark diminishes the antiscreening effect of the surrounding virtual gluons, so the contribution of this effect would be to weaken the effective charge with decreasing distance."

Don't worry, that's as technical as we will get, I just want you to get familiarized with quantum speak. The Non-Abaelian field gauge theory applied to quarks is called Quantum Chromodynamics (QCD), is the current major accepted theory for strong interactions, and is being experimentally proven at an accelerator near you. Welcome to the world of quantum's charge, fields and virtual particles!

The Standard Model…unifies all the interactions under a common model where all particles are considered together: The weak interactions mediated by electrons, muons, taus and neutrinos; the quarks are both strong and weak interactions mediating the forces for electromagnetism; the strong force is mediated by the massless particles the photon and the gluon; and for the weak force, the W and Z particles, which do have mass.

Coupling strength decreases with a decrease in energy, which is to say, that in terms of quantum mechanics, it increases with distance. This is why you see the little spring in the diagrams holding quarks together, as they liken this force to the action of a spring. Photons explain Coulomb's law for large distances and gluons explain the binding force between quarks in the hadrons.

The Standard Model ignores the gravitational force because it is so small compared to the others at the elementary particle level as to be inconsequential. What is the quantum version of gravity that works at the tiny distances? Why can't the gravity field be quantized into gravitons like the magnetic field is quantized into photons? Einstein's gravity is not renormalisable. Feynman diagrams don't work here.

Which theory is incomplete?

The Standard model shows unification of the electromagnetic and weak forces at energies above 100 GeV and the strong force unifies with the weak force at above 10^{15} GeV. Mathematically, at least, it can be shown that at energy level above 10^{19} GeV gravitational force is as strong as the other forces. The only time that could have happened is at 10^{-42} seconds after the Big Bang.

The Superstring Model is just such a proposal that claims to unify all the forces including gravity.

With one dimensional strings Einstein's theory for low energies are compatible with the Standard Model at the energy levels already achieved. Being a finite quantum theory, Superstring Theory is consistent with a perturbation theory for gravity. Additionally, the model unites force particles with matter particles.

Or, so we have been told.

Chapter Three - What is Force Anyways?

In the previous two chapters we discussed the three forces of the Standard Model of Physics: The electromagnetic force and the strong and weak nuclear forces. We covered these with broad strokes to familiarize our selves with the terms, and to get a general sense of what Quantum is all about and where it is coming from. We discovered that the history of the various theories of forces extends from at least Isaac Newton's Law of Universal Gravitation to current day quantum chromodynamics and Superstring Theory.

We also learned that scientists generally ignore gravitational force at the subatomic level since that force is magnitudes smaller than all the others and inconsequential as far as their calculations are concerned. We noted that there is a disconnect between Einstein's General Theory of Relativity and Quantum Physics (except for QED), and that scientists are trying to reconcile the two with a theory of quantum gravity in order to unite the forces of the Standard Model with gravitation.

Perhaps this is why some sources refer to the three forces of the Standard Model and others the four forces of the Standard Model:

Berkeley University's website, "Particle Adventure," which is supported by the Department of Energy (DOE) and the National Science Foundation (NSF) says: "The Standard Model is a good theory…..but… gravity is not included in the Standard Model."

CERN: "The Standard Model explains how the basic building blocks of matter interact, governed by four fundamental forces." https://home.cern/about/physics/standard-model

At first I thought maybe someone isn't getting their memos, then, I thought it is, as Nature says in reference to CERN's LHC experiments, "Researchers are desperate to unravel physics beyond the standard model, a description of particles and their interactions. But physicists know that the standard model is not the end of the story, not least because it doesn't include gravity or explain dark matter."

https://www.nature.com/news/rare-particle-decays-offer-hope-of-new-physics-1.22103

If you read on in the CERN article about the Standard Model you'll find this caveat:

"However, the most familiar force in our everyday lives, gravity, is not part of the Standard Model, as fitting gravity comfortably into this framework has proved to be a difficult challenge."

Rest assured, whether Hubble will find Dark Matter, or Fermilab's NuMI or DUNE projects crack Father Universe's neutrino code, a New Science and a New Standard Model will be forthcoming.

Modern physics is very weak at scientifically defining their Key Terms and this leads to different interpretations among scientists and confusion among physics students and teachers alike. So far, we have been told a lot about "force" but no one defines this crucial term which makes or breaks their presentations.

Here are four definitions for the Key Term force:

Dictionary.com says this: "In physics, something that causes a change in the motion of an object."

Wikipedia's definition is similar: "In physics, a force is any interaction that, when unopposed, will change the motion of an object. Force can also be described intuitively as a push or a pull."

We find this definition at the physics classroom: "A force is a push or pull upon an object resulting from the object's interaction with another object. Whenever there is an interaction between two objects, there is a force upon each of the objects. ... Forces only exist as a result of an interaction."

Finally, this from the on line science dictionary Wolfram: "The physical influence that produces a change in a physical property."

Force, then, is some"thing" that causes a change in motion, an interaction when unopposed that changes the motion of an object, a push or a pull caused by the interaction between two objects or a physical influence that changes a physical property. That should cover all the bases!

To be clear, CERN has this to say about "force":

"CERN researchers confirm existence of the force. Physicists at the European Laboratory for Particle Physics announced today that an invisible Force permeates the universe, binding the galaxy together."

Before we dig deeper into the three...er four forces, let's see if three of these things is like the others but one of these four is not. Does Newton's gravitational force conform to any of these definitions? Britannica's on-line definition reads very closely to many other sources and goes like this:

"Newton's law of gravitation; statement that any particle of matter in the universe attracts any other with a force varying directly as the product of the masses and inversely as the square of the distance between them." Wiki says pretty much the same thing with this variation at the end, "the distance between their centers."

Other variations read, "every mass attracts every other mass," or they replace "mass" with "object," or body," and occasionally, "point mass."

So, you're probably wondering, "What did Newton actually say?" Newton's book, "Philosophiæ Naturalis Principia Mathematica" is at Cambridge University and images of every page are on-line for your perusal. It's in the original language of Newton, but a translated version can be found here:

https://archive.org/stream/newtonspmathema00newtrich/newtonspmathema00newtrich_djvu.txt

So what did Newton actually say?

"The principle of Universal Gravitation, namely, that every particle of matter is attracted by, or gravitates to, every other particle of matter, with a force inversely proportional to the squares of their distances is the discovery which characterizes The PRINCIPIA."

It seems as though Newton is saying that gravitation is a force, after all, he uses the term when he states that it is the attraction between particles. We'll return to gravity with more detail in another chapter. Meanwhile…

Is Electromagnetism a Force?

For more details on electromagnetism including Maxwell's, Kirchoff's and Coulomb's Laws read Rational Science Vol. 5, Chapter Thirty Two – Plasma the Fourth State of Matter?, Chapter Thirty Three – Ions, Charge and Matter, Chapter Thirty Four – Elementary Charge, Chapter Thirty Five – Batteries, Current Flow and Ions.

If the mediator of electromagnetic force is the electromagnetic field, as we have been told, then we should understand what is meant by "field." Let's accept that some thing mediates force. It makes sense, that, when I push you, the force "push" is mediated by me against you. Or, more accurately, there is an interaction between two bodies.

Field can be many things in ordinary language. As a transitive verb, the secretary can field calls. The farmer can mow the field (noun). A person can conduct a field study (adjective). Field can be a countable noun as in field of vision or a place in a computer program where you enter data.

However, in science, there are only two categories in which our Key Terms can fit. A term is either an object or it is a concept. Collins dictionary says this about "field" as it relates to physics: "countable noun - A magnetic, gravitational, or electric field is the area in which that particular force is strong enough to have an effect."

When I was studying for my degree in electrical engineering we talked about electric fields being caused by charged particles. Charged particles produce a force field around them and these fields can interact with other fields. Magnetic and electric fields travel at right angles to each other. We can't have one without the other. Why? Why do magnets attract in one direction and repel in another? What is negative or positive charge? Which way does electricity flow; from negative to positive, or positive to negative? I received my degree, but I never found out the answers to these questions. And no one to this day can answer them. At least particle physics doesn't have the answers. They never will if they don't find the objects behind the phenomena. That is, the physical mechanism(s) responsible. And they won't find the objects because they are not looking for them. They are turning concepts into the objects themselves.

Four hundred years of science and scientists still can't explain the workings of a simple magnet. Listen to Richard Feynman when he is asked about why magnets attract:

"You have to know what it is that you're permitted to understand and allowed to be understood and known and what it is you're not. That's just one of things you'll just have to take... is electromagnetic repulsion and attraction. I can't explain that attraction in terms of anything else that is familiar to you. If I told you the attraction was like rubber bands and you asked me about rubber bands, I would have to explain it to you with the very thing electromagnetic attraction which I am trying to explain to you. You would be cheated very badly, you see, so I'm not going to be able to give you an answer to why magnets attract each other except to tell you that they do. And to tell you that is one of the types of forces in the world, there are electrical forces, magnetic forces, gravitational forces, and others, and those are some of the parts. I could go further and tell you that the magnetic forces are related to the electric forces very intimately. That there are relationships between the gravity forces and the electrical forces remains unknown. But I can't do a good job...any job of explaining magnetic force in terms of something else that you're more familiar with, because I don't understand it in terms of anything else that you're more familiar with."

If a field is an area where charged particles are found then how can fields interact? They can't. Only charged particles can interact. But this is not what they are saying. How can an area contain a force which has an effect? It can't. Particles may interact with particles by pushing and pulling on them. But this is not what they are saying. When challenged, the theorist will use the definition for field that best fits his presentation at the instant moment.

An electric field is defined as the electric force per unit charge. Electric charge is a property of a unit of matter telling us if it has more or fewer electrons than protons compared to surrounding atoms. As well, an electron or a proton both carry elementary charge which are equal but opposite (negative and positive).

Huh? Charge is denoted by the difference between the number of an atom's electrons or protons and those of surrounding atoms, but it is also the difference between a proton and an electron which "carry" either a negative or positive charge. How can an electron or proton carry charge (a ratio of electrons and protons)? Obviously, they mean something different by elementary charge. I guess it means they carry a negative or positive sign. See Rational Science vol. V; Chapter Thirty Four – Elementary Charge

So an electric field, then, is the measure of some kind of interaction (force) between atoms based on their number of electrons and protons. But I thought a field was a location. Now we find that it is also the interaction between units of matter. But fields are also considered objects that can interact with other objects, and also mathematically a region where point particles can be located.

"Electromagnetic forces occur when an electromagnetic field interacts with electrically charged particles. The formula that describes the behavior of electric and magnetic fields, and their interactions with matter, were derived by Oliver Heaviside (1831-1925), but are now called Maxwell's equations after James Clerk Maxwell (1831–1879)." - Plasma universe

So, we see concepts interacting with concepts, but it gets far worse once you get into quantum field theory. Listen to quantum physics buff Viktor Toth who says, "Elementary particles are not classical objects. They are not teeny-weeny balls flying about in empty space. They are excitations of quantum fields, which are only localized under special circumstances (e.g., when we do a measurement) which is when they appear particle-like."

Quantum physicist Matt Strassler says: "We should remember that electrons, like all elementary "particles," are really "quanta," tiny ripples in quantum fields."

To complicate things he adds that there are also virtual particles:

"The force exerted by the positron on the electron can cause the electron to become, sometimes and briefly, a virtual electron and a virtual photon. Virtual 'particles' aren't real particles; a real particle is a well-behaved ripple in quantum fields, but a virtual particle is a more generalized disturbance of those fields."

Hmmm? Well, whether "real" or "virtual," any phenomenon requires surface to surface contact between at least two (or more) objects. No wonder it is as clear as dirty dishwater what mediates electric charge or force or why Richard can't tell us how magnets work!

Finally, here's - Alan Bustany, Physics Student, who sums things up quite nicely.

"The Standard Model of particle physics rather has atoms as an emergent property of the various quantised fields that exist. The precise way in which these fields interact with space-time at the extremes of small distance (the Planck length much smaller than an atom) and large density (the singularity thought to be within a Black Hole) are not understood, and await a Grand Unified Theory (GUT) or Theory of Everything (ToE). Nevertheless both Quantum Theory and the General Theory of Relativity are the most accurate theories we have ever had, despite their shortcomings!"

That's right, Alan, they are not understood! How real or virtual atoms don't really exist but are only properties which emerge from fields that do exist remains to be understood. AND it will never be understood. Until the phiz whiz sorts out the objects from the concepts, he or she will never understand anything.

Chapter Four - Quantum Fluctuation

Alan summed up our last chapter by informing us that atoms emerge from fields but that physicists don't understand how these fields interact with spacetime at small distances; QM and relativity are still the most accurate theories that they have.

They are "accurate," because most of the formulas and equations work out without a lot of fudging, but they don't understand them because they have no physical interpretations for their calculations. Quantum mechanics and Relativity have their "shortcomings" because space time, Planck units and black holes are irrational objects. An abstract object, like a string with length but no height or width, is an irrational object. Black holes contain an infinitely dense point-mass singularity and are therefore impossible objects. It is because these abstractions are derived by mathematical reification and described using unambiguously defined terms like point and mass, that they do not correspond with reality.

We saw that by interchanging terms such as particle, mass, object or point particle, a field can take on different meanings. It certainly means something different to interpret a field as an area where particles interact as opposed to a geometric region where point particles are located. Now place charges in that grid and we have a higher level of abstraction and irrationality. Moving electric fields create magnetic fields, fields interact with objects and other fields, and fields even interact with spacetime. And when we get to the highest level of quantum abstraction, terms continue to fluctuate at an even higher rate. I call this Quantum Fluctuation. Of course, this term is already being used by the mechanics and is used to describe a fantasy change in *vacuum point energy* allowing the production of virtual particle-antiparticle pairs. Yet another thing we can thank the Heisenberg Uncertainty Principle for. Being uncertain is a great excuse to imagine all kinds of stuff! *Vacuum point energy or zero point energy is covered in Rational Science VOl. II, Chapter Twenty Eight – Zero Point Energy Nonsense*

Strong Force

We only briefly mentioned the Strong Force, and we used a couple of terms, coupling constant and perturbation expansion, to sound like we understood what we were talking about. We really don't. No one does. In the words of Richard Feynman, "I think I can safely say no one understands quantum mechanics." Coupling constant is a parameter; a number that tells us the strength of force in a particular interaction. Perturbation expansion is a mathematical theory for determining an approximate solution for a problem. We're really not interested in the math at this point, we just want to see if we can arrive at some physical interpretations of what this math is all about. And, btw, I am really only using "we" because it has a more authoritarian sound to it than "I."

Quarks are the mediator of the strong force that hold things together in the nucleus of an atom. As in all particle theories, we need to understand the underlying field theory of quarks. Quantum field theory is a combination of quantum mechanics and relativity and hence a mathematical system used to describe the behavior of sub atomic particles. Quantum electrodynamics is used to describe the interaction between the EM force and charged particles, and Quantum chromodynamics describes the interactions between quarks and the strong force.

Quarks come from the family of fermions which include leptons and some more familiar particles; protons and neutrons. According to the Standard Model, fermions are among the 12 "building blocks" of matter. The five bosons are the mediators of interaction. Clearly then, particle physicists have delineated matter from mediators. QCD predicts that it is the gluon which exchanges the force between quarks that bind protons and neutrons in the atom. Yet, if quarks are the particles that hold atomic nuclei together, then what are the particles that mediate force between fermions and bosons, in this case the quark and the gluon?

Gluons exchange the strong force between quarks like photons exchange the electromagnetic force between charges particles. The gluon is not the force itself (fwew! For a minute there I thought we were going to have interaction at a distance!), it is the mediator of the strong force that holds quarks together, and quarks hold fermions together.

"Statistically," fermions can not occupy the same place as other fermions but bosons can. A photon is a boson, which explains why two flashlight beams can cross through each other without any interaction between photons. Wait! Hold yer horse Gabby! Them there's statistics, and we all know there's lies, damn lies and statistics!

There is something called gamma-gamma physics which is all about the interaction between photons. It seems that passing through each other slows photons down. This can happen at very high energies in an optical device, or light can scatter light "in space." AND, I say AND above a certain threshold, energy is converted to matter! Let's take a moment to define matter as it is meant here.

Matter: any substance that has mass and takes up space by having volume.

So bosons, like the photon, which have no mass and do not take up space, can collide with enough energy to create matter. In other words, something without mass can create something with mass. Creation physics at its finest.

Hold it right there AGAIN, pardner. You said quarks, which have mass, are the mediators of the strong force. Then you said gluons are massless particles that exchange (mediate) the strong force between quarks. So we have massless particles which interact with particles that do have mass and these interactions allow the strong force to bind quarks to the atom. Sounds pretty mysterious to me!

Now, don't pick fly shit out of pepper, we're using the magic words of force and mass and energy here. Take that and Werner's principle and we can do anything! If you don't believe me, get a load of this. An article at Duke.edu said, speaking of the top quark:

"Only the Tevatron at Fermilab has enough energy to produce such a massive particle, which is incredibly huge for a "subatomic" particle; more massive than a total gold atom."

Hundreds of physicists share a Nobel Prize in physics for this, so don't question me, ya' hear! All atoms have one of these top quarks and the top quark is hundreds of times more massive than the entire gold atom. Let that sink in. A gold atom is an apple. It has three seeds. One of those seeds is hundreds of times more massive than the apple. Kewl, eh?

What? Now I didn't just fall off of the apple truck, mister. The seed is heavier than the apple?

Chemicool.com says mass is a measure of the amount of matter in an object. Mass is usually measured in grams (g) or kilograms (kg). Mass measures the quantity of matter.

No, I didn't mean it that way, silly. I meant it the way techtarget.com means it:

"Mass (symbolized m) is a dimensionless quantity representing the amount of matter in a particle or object."

Wish those guys would get together and come up with a scientific definition! One that can be used consistently.

Chapter Five - The Weak Nuclear Force

Gravity and electromagnetism operate over long distances compared to the strong and weak nuclear forces which are confined to distances within the diameter of atomic nuclei. The weak force is thought to be responsible for beta decay and the strong force, as we learned, holds things together in the nucleus of the atom. While the exchange particle for electromagnetism, the photon, is "well known," the predicted particle of exchange for gravity, the graviton, is yet to be found. We'll cover gravity in the next chapter, but for now let's concentrate on the weak nuclear force.

Understanding of all the forces is based on field theories for each of them, but, particle physicists hope to unify all the forces under one grand unifying field theory. Until then they'll have to be satisfied with a unification of electromagnetism and the weak nuclear force called the electroweak theory.

All forces interact in certain ways with their particles because of that particle's properties. Whereas electromagnetism has charge and the strong force has color, the weak force has flavor. As you recall, electromagnetism has the boson "photon" and the strong force has bosons known as "gluons," the weak force has two bosons; the W and Z bosons. These are known as the carriers of the weak force. "Flavors" means types, and the different flavors interact "weakly" as W and Z bosons are exchanged between them.

Among their duties as sub atomic particles, W bosons are responsible for neutrons decaying into protons. Z bosons carry a neutral charge. Well, let's just let CERN tell the story:

"By emitting an electrically charged W boson, the weak force can cause a particle such as the proton to change its charge by changing the flavour of its quarks. In 1958, Sidney Bludman suggested that there might be another arm of the weak force, the so-called 'weak neutral current,' mediated by an uncharged partner of the W bosons, which later became known as the Z boson."

Like all elementary particles except the Higgs boson, W and Z bosons have something called spin. Not spin like Lebron James spins a basketball on his finger, because elementary particles have no top, bottom or sides (they are mathematical points), but a spin number (1) which represents a measurable quantity that uses the same units of measurement as angular momentum. Particles with spin have different numbers depending on whether they correspond to a vector field, scalar field or a tensor field.

W and Z bosons have a spin number of 1 because they are force carrying bosons that correspond to vector fields, in other words, they have direction. Higgs is 0 because it corresponds to a scalar field which has no direction, and if they ever find the graviton, it will be a 2 because it corresponds to a tensor field which warps space. Got it? Good!

Wiki tells us the "weak force," or the "weak interaction" or the "weak nuclear force" is the "mechanism of interaction between sub-atomic particles that causes radioactive decay." Because other forces have their own titles, like the strong interaction has Quantum chromodynamics (QCD) and the electromagnetic force has Quantum electrodynamics, the weak nuclear force chose the title, "Quantum flavordynamics. (QFD)" Unfortunately, that one hasn't stuck as this force is more widely know by the Electro-weak Theory (EWT). What's in a name anyways? A weak force by any other name is still weak.

Or is it? No! The weak force is actually a very strong force (it's stronger than gravity), but since the particles (W and Z) are so big (or is that massive?) and the force only acts at such small distances, it's called weak.

Fermilab's website tells us that Beta decay is but one example of the weak force, and it's my personal favorite, so let's talk about that. Beta decay causes the neutron of an atom to disappear and be replaced by a proton, an electron and an anti-electron (neutrino). So the weak force, which is strong, is responsible for annihilating atoms and replacing them with a new atom and a new electron which is really not an electron. This shape shifting occurs by way of a down quark disappearing and an up quark taking its place. Soon, however, this magical quark becomes an electron and a neutrino. * This weak force is so strong, without it the sun would cease to exist, and some matter, like uranium and plutonium would become stable. Why, you could let your toddlers play with it! I know it sounds far-fetched, but if you just look at Fermilab's diagram, you'll see why this is possible.

*In beta minus decay that's an anti-neutrino…er…anti-electron neutrino. Don't confuse this with neutrinos and the other antineutrinos the muon and tau antineutrino. AND don't be confused that the neutrino is its own antineutrino. Let Berkley Lab straighten us out:

"What's the difference between neutrinos and antineutrinos? Not much, it seems: neutrinos always spin left and antineutrinos always spin right. 'Spin' isn't Newtonian rotation, however, it is a quantum number. In the case of neutrinos and other leptons, the spin number is one-half, which may sound non-common-sensical but is nevertheless deep – not to mention that it's a useful way to tell otherwise indistinguishable particles apart."

The Lab continues in our edification:

"Fusion reactions like those in the sun involve so-called beta-plus decays: a proton changes to a neutron plus a positron, the electron's antiparticle, and an (ordinary) neutrino. To conserve angular momentum and balance the quantum variables – including those invoked by the weak force, which governs neutrinos – ordinary neutrino spin has to be left-handed. Fission reactions like those in a nuclear power plant involve beta-minus decays: a proton changes (again mediated by the weak force) to a neutron plus an electron and an antineutrino – whose spin must be right-handed. The possibility arises that left-handed neutrinos and right-handed antineutrinos with even a little mass are actually the same – or, to put it another way, that the neutrino is its own antiparticle."

That clears a lot up. Anyways, when a muon changes into an electron, it has not disappeared and been replaced, it has merely changed its flavor. Well, now, that is much easier to take. It's like a spoonful of sugar helping the medicine go down! Thank you Doctor Particle Physicist.

Now you understand how the sun converts hydrogen into helium. Uh-huh.

You are probably wondering what the difference is between the bosons W and Z. W's are electrically charged. Either –W or +W. This is the particle that changes other particles. The weak force uses a W to change the flavor of a quark, which in turn changes a neutron into a proton or a proton into a neutron. The Ws, in this manner, set the suns a blazing.

Zs carry a neutral electrical charge and were predicted to be massless along with the Ws, but smart particle physicists knew that W had to be a heavy weight in order to be a force carrier at such short distances. Enter the Higgs mechanism. Seek and ye shall find…so… CERN observed something "consistent in appearance" to this boson in 2012.

From the Annenberg Learner brought to you by Harvard/Smithsonian Center for Astrophysics

"The way the Higgs field gives masses to the W and Z particles, and all other fundamental particles of the Standard Model (the Higgs mechanism), is subtle. The Higgs field—which like all fields lives everywhere in space—is in a different phase than other fields in the Standard Model. Because the Higgs field interacts with nearly all other particles, and the Higgs field affects the vacuum, the space (vacuum) particles travel through affects them in a dramatic way: It gives them mass. The bigger the coupling between a particle and the Higgs, the bigger the effect, and thus the bigger the particle's mass."

https://www.learner.org/courses/physics/unit/text.html?unit=2&secNum=6

The electroweak theory combines four force carriers and two uncharged particles that mix together and evolve into each other as they travel from point A to point B (just like the other shape shifting particles, the electron, tau and muon neutrinos.) To delve deeper into the mystical world of Quantum magic, one needs to look at parity, charge conjugation and symmetry breaking along with many other irrational ad hoc mechanisms. We'll discuss this more in a future chapter.

Chapter Six - Is Gravity a Force?

According to NASA, "We don't really know.... if we are to be honest, we do not know what gravity "is" in any fundamental way - we only know how it behaves.

"Gravity is a force of attraction that exists between any two masses, any two bodies, any two particles."

https://starchild.gsfc.nasa.gov/docs/StarChild/questions/question30.html

Really? Is gravity a force?

Not according to Einstein. He said that gravity was a result of mass curving spacetime.

"So, to summarize, general relativity says that matter bends spacetime, and the effect of that bending of spacetime is to create a generalized kind of force that acts on objects. However, it isn't a force as such that acts on the object, but rather just the object following its geodesic path through spacetime." - Dr Jolyon Bloomfield

http://curious.astro.cornell.edu/physics/140-physics/the-theory-of-relativity/general-relativity/1059-if-gravity-isn-t-a-force-how-does-it-accelerate-objects-advanced

How We Know Gravity is Not (Just) a Force

"When we think of gravity, we typically think of it as a force between masses. Forces are easy to understand as pushes and pulls.

"In general relativity, gravity is not a force between masses. Instead gravity is an effect of the warping of space and time in the presence of mass. "

https://physics.stackexchange.com/questions/61899/why-do-we-still-need-to-think-of-gravity-as-a-force

Let's take a look at a few Quora answers.

Are you sure gravity is a force?

"This answer is actually going to be a bit philosophical, because it really depends on what a force is. Here are several ways to think about it." –Jane Jacobs, B.S. Applied Mathematics, University of Colorado Boulder

Mathemagician Jane goes on to give three interpretations. You drop a brick on your foot and it hurts, so it must be a force, right? You can interpret gravity geometrically using the Riemann tensor, or you can say the brick just took the straightest path through space time curvature, and is therefore not a force. Then he says this:

"What I'm now going to try to do is persuade you that it really doesn't matter which interpretation you take, as long as you use the same equations."

Jen Jameson, Developed Inertial Guidance Systems:

"Yes - gravity is due to a force but it may or may not be directly acting upon the object in question."

Jen tells us that although relativity is a geometric construct... "What is sometimes overlooked is that the postulated bending of space by matter itself requires a force. That force is present in the Einstein's equation as it was also in Newton's formulation $F = MG/r^2$."

So, it seems, we still are having problems with that force thang. NASA says we don't know what gravity is, but that it is a force. Ask a handful of physicists and you get a handful of different answers. No, it is not a force, it is a force, it doesn't matter if it is a force or not as long as the equations work out, and it isn't a force per se but it requires force to move objects.

If gravity isn't a force, how does it move objects?

Let's be accurate. The sun doesn't pull on the earth, and the earth doesn't pull on the moon, it isn't the force "pull" it is something called acceleration. When the sun weighs down the space tarp the earth rolls down it and accelerates.

Relativity says that "the energy of a mass tells spacetime how to bend, and the bent space tells energy how to move." The paths that objects take are called geodesics. The earth's orbit around the sun is a result of it rolling around the sun's gravity well.

As we have discussed before, gravity is not part of the Standard Model of physics but wants very much to join the other Forces and make it The Fantastic Four.

Theoretical physicists make many claims and scientists do experiments which, as far as they are concerned, validate them. Eddington observed an eclipse and compared it to other eclipses noting that the position of stars had shifted when the sun was nearby. The amount of shifting was mathematically consistent with Einstein's theories, but not Newton's.

Sky gazers observe something similar with galaxies and quasars and they call this gravitational lensing. Their measurements are used to estimate masses of galaxies and stars, and to see dark matter's effects.

Time delay experiments within the solar system are also "explained" by the Shapiro effect (and others). Massive celestial bodies like the sun warp space causing light to travel a further distance as it follows a curved path to get to us.

The motion of binary stars, we are told, can cause ripples in spacetime so that gravitational waves spreading out from the system carry away energy and as a result the stars move closer together. This means a reduced orbital period and the effect is called inspiralling.

Astronomers imagine that pulsars' pulse rates vary due to this effect, and so are convinced by these many observations that mass warps spacetime.

Close to the earth the effect of frame dragging becomes apparent. Not only does the mass of the earth weigh down the space fabric, its rotation twists and can be measured due to the Lense-Thirring effect. NASA put a satellite in space (Gravity Probe B) with a spherical gyroscope and confirmed frame dragging due to the precessing (a change of orientation over time) of the gyro.

All these measurements and experiments convince the mathemagician, sky gazer and lab monkeys that gravity is an effect of spacetime. They assure themselves that the shape of "the Universe" has gravity built right into it.

If relativity predicts all these observations, and is therefore confirmed, why are the theoreticians still trying to unite relativity with quantum?

Well, Newtonian physics is incomplete. It couldn't explain the precession of Mercury like Einstein's equations do. Sure it's still very useful here on earth, and we can accurately launch rockets to Pluto with it, but relativity predicts so much more like expanding universe, dark matter, dark energy and what side of the bed you are going to get up on in the morning. Similarly, relativity can predict a lot about the movement of large celestial bodies, but the equations are not very useful at micro scales. Tiny things behave differently. Quantum, on the other hand, is very accurate in its predictions at the atomic and subatomic level.

Chapter Seven - What is Quantum Gravity?

Quantum gravity: "In theories of quantum gravity, the graviton is the hypothetical elementary particle that mediates the force of gravity." – WIKI

Huh? Wait! I thought that gravity wasn't a force. AND what's with the particle search if gravity is a result of warped spacetime?

What is Gravity?

Maybe Don Lincoln can fill us in. After all he's a particle physicist at CERN and Fermilab. He has authored over 500 scientific papers, is an adjunct professor at the University of Notre Dame and has given lectures all over the world, made YouTube videos, and written articles for Fermilab Today. Don has also written a book, "The Large Hadron Collider: The Extraordinary Story of the Higgs Boson and Other Stuff That Will Blow Your Mind."

Don tells us, historically, the first real study of gravity is often said to have begun with Isaac Newton. As a result we have Newton's second Law of Motion $F=ma$ and the law of gravitation $F = G(m_1 m_2/r^2)$. Although Newton could describe acceleration and gravity very well with his equations, as we learned in a previous chapter, he never understood the underlying physical mediator of gravitational force.

Then, along came Albert Einstein with his warped space and time. He claimed that it was not far reaching fields that mediated the force of gravity and that gravity was NOT a force. It was the "gravity well" formed in spacetime by massive bodies that is responsible for gravity. The sun, for example, pushes down on spacetime and causes smaller bodies like the earth to roll around it like a ball on a roulette wheel. That's right, Einstein used gravity to explain gravity! It may look like the earth is orbiting the sun in an ellipse, but actually it is following a straight path through warped spacetime.

Whereas Newton's Laws work very well here on earth and can even get a small object like a rocket to the planet Mercury, Einstein's equations predicted the unusual orbit of Mercury with impressive accuracy, and so got the attention of the scientific community as being better equipped for "explaining" how large objects act.

Though Einstein's General Relativity was excellent for explaining massive objects, physicists were "discovering" a micro world of tiny discrete particles mediating the forces of electromagnetism (EM) and "the" strong and weak forces.

What Are Gravitons, Really?

A whole new magical world opened up with force-carrying particles like the massless photon, responsible for EM forces; weak nuclear force created by moving W and Z boson particles; and the strong nuclear force being transferred by gluons. Naturally the Phiz Whiz asked him or herself, "Is gravity mediated similarly?" Hence, the search for the ever illusive graviton.

The nudnicks at CERN and Fermilab and in laboratories elsewhere will search relentlessly for this elusive particle of gravity. Their excuse for not finding it is that it is so small, and of course massless, that it is more difficult to catch than a Pegasus.

The range of the force due to gravity is without limit- yes infinite, but the force of gravity weakens over very, very small range, that is $1/r^2$. Therefore scientists know that the graviton particle must be massless like the photon. If it was NOT massless, the little 2 in distance squared would be different. That can't be the case because lab rats have measured this with incredible accuracy. This also means that the graviton would move at the speed of a photon. Yep, the speed of light.

General relativity is a geometric theory and spreads mass and energy around the universe in a "rank two" tensor, that is, a four by four matrix. The graviton, then, is a spin 2 particle according to quantum mechanics.

Why Haven't We Found Any Gravitons?

Because gravity is so weak and gravitons are so frail and interact so weakly. Why does the apple fall down instead of up into the sky? Because there are so many gravitons being given off by the earth!

But do not be faint of heart, scientists think there could be several kinds of gravitons and some of these could be detectable. Since gravity COULD have access to other dimensions, it is possible that gravity is much stronger than we formerly thought. So gravity is not the weakest force after all, it is just spread out over more dimensions.

Since particles are also waves, the graviton could be vibrating in this other dimension. Of course it would require a special kind of vibration, that is, a wavelength in integers only (in order to fit evenly in this dimension). One way that these wavelengths could fit into the small extra dimensions could be imagined as a sine wave wrapped around a cylinder. Each of these different integer wavelengths could be different types of gravitons with some of these having mass. So, wala, even though these are thousands of times smaller than a proton, we should be able to detect them! Of course, these tiny particles would still not have mass at the larger scale of classical quantum mechanics theory.

SUMMARY

Let's summarize what we have been told by authority Don Lincoln:

Newtonian gravity works fine on small bodies. Einsteinian gravity works great for large bodies, but we need Quantum gravity for the very small itty bitty tiny micro world below the atom in size.

According to Newton:

F = ma
F = G m1m2/r^2

According to relativity:

Gravity isn't a force.
Gravity explains gravity.

According to quantum:

Gravity will blow your mind.

Force carrying particles transfer all the forces including that of gravity.

Graviton particles are massless and travel at the speed of light.

Graviton particles also have mass at scales thousands of times smaller than the proton.

Graviton particles are also waves that vibrate in other dimensions.

Chapter Eight - Evidence for Quantum Gravity

In previous chapters we assumed many things in order to accommodate the quantum magician. In this and following chapters I will highlight their Key Terms and phrases then define them or point out the irrationality and also refer you to sources where you may study these in greater depth.

What Evidence is there for Gravitons?

Evidence is your opinion, extra scientific, and therefore NOT part of our scientific inquiry. We observe something, it makes us curious as to the physical mechanism behind the phenomenon, and THEN we apply a rational scientific method to understand and explain it. The hypothesis illustrates the objects, the Key Terms are defined scientifically, and the assumptions are presented for explanation in the theory. What you believe, is YOUR conclusion and yours alone. Science is done! Whether or not you were convinced, considered something as evidence, or take a counter position is of no concern to anyone else. To discuss with the luxury of detail go to the Facebook Group "Rational Scientific Method." You may also refer to my books on rational science.

Rational Science Vol. II; Chapter One – The Rational Scientific Method, Chapter Two – Scientist, Science, & The Scientific Method, Chapter Three – Scientific Method? For Dummies!, Chapter Four – Hypothesis, Theory, Conclusion, Chapter Five – Science & Technology - Conceptual & Empirical, Chapter Six – Experiments Are They Part of the Scientific Method?, Chapter Seven – Pseudo-Scientist Index, Chapter Eight – Proof Is For Alcohol

Here, for sake of their argument, let's see what the mechanics are calling evidence.

Let's take a brief look at Brian Koberlein's article "Gravity." What is his evidence? Black holes and gravity waves!

"During the Archean Eon of planet Earth, when life was figuring out how to harness energy from the Sun, two black holes in a distant galaxy merged with a ripple of gravitational waves. Over the next 2.9 billion years these ripples traversed a vast and empty space, while on Earth a plucky little species of bipeds learned to use lasers and mirrors to measure gravitational vibrations smaller than the nucleus of an atom. When the gravitational ripples reach Earth, they become humanity's third detection of merging black holes."

Gravity is the phenomenon of interest, and waves are what physical mediators are doing. The ocean waves; the amber waves of grain; I wave my hand, and so on. In physics, we use physical objects to explain phenomena, because we understand that all phenomena are the result of surface to surface contact between two or more objects. We never move concepts around. "Gravitational waves" is moving a concept.

Gravitational; adjective: denoting a forceful attraction or movement toward something.

Wave; verb: In physics, a wave is a disturbance that transfers energy through matter or space

Furthermore, in physics, we use adjectives to qualify nouns. What can "attractive disturbance" even mean?

Give LIGO credit for "detecting" these tiny gravitation waves - evidence of the merger of black holes. Since the "newly" consolidated black hole is so very far away, Brian and his cohorts... er... colleagues think that this will allow them to "test Einstein's theory in new ways, particularly a quantum aspect known as gravitons."

Rational Science Vol. I; Chapter Twenty Two – Black Holes, Rational Science Vol. II, Chapter Thirty Four – Gravitational Lensing and the CMB, Chapter Thirty Eight – Gravity ...Well?, Rational Scientific Method Vol. V, Chapter Ten – LIGO and Gravitational Waves, Rope Hypothesis and Thread Theory, Chapter Thirty Six - Gravity Basics, Chapter Thirty Seven - Big G in Newton's Law of Gravitation, Chapter Thirty Eight - Big G Two, Chapter Thirty Nine - Tension as Numbers

Next, Brian tells us that, "Gravitons are the only field quanta we haven't observed. Gravity is the weakest of the four fundamental forces, so to directly observe a graviton you'd need something like a Jupiter-mass detector orbiting a neutron star."

Popular Mechanics magazine, May 2018, said this: "The biggest roadblock for physicists is that the graviton, if it exists, is so weak that it's pretty much impossible to detect. In order to spot it, we would need a detector so massive it would collapse into a black hole."

Observation is what leads us to the scientific method. Reality does not depend on human perception or observation. It is because the human senses are limited and flawed that science must be as objective as possible. The scientific method is observer independent. One must apply rationality, reasoning, and critical thought at the conceptual stage in the hypothesis. There is no way to conceive of an infinitely dense massless singularity, or black hole. It violates reason, logic and basic math. One can not divide by zero as is needed in the density formula applied to a black hole.

Exist, is the Key Term for all of physics. Yet, it is never defined. Ask a mathematician to define exist and he will send you down the hall to philosophy. Apply this useful definition of exist and see why their entire house of cards comes falling down:

Exist – object with location in respect to all other objects; something somewhere, physically present.

Rational Science VOl III; Chapter Twenty Seven – Black Holes, Chapter Eleven – Existence

Brian continues, "One of the key predictions of relativity is that gravitons should be massless, like photons. As a result, gravitational waves should always propagate at the speed of light. With this new merger we can test this idea through a property known as dispersion. Dispersion occurs when waves originating from the same source travel at different speeds. You can see this in a prism, where sunlight is spread out into a rainbow of colors.

This is caused by the fact that the speed of light through glass varies with wavelength or color. "

There are a number of problems with this, not the least of which is that science does not predict, it explains. Predictions and observations are opinions and are extra-scientific. If I observe an apple fall a few times and measure the speed and distance traveled, I can "predict" how fast an apple falls. What does that tell me? It does not tell me when an apple is going to fall. Now THAT would be a real prediction. Something that already happened, a consummated event, is described and should then be explained. Something that we have observed happen repeatedly can lead us to think that there is a high degree of probability that it will happen that way again. But that is not really a prediction - it's an educated guess.

Rational Science VOl III; Chapter One – The Rational Scientific Method, Chapter Two – Scientist, Science, & The Scientific Method, Chapter Three – Scientific Method? For Dummies!, Chapter Four – Hypothesis, Theory, Conclusion, Chapter Five – Science & Technology - Conceptual & Empirical, Chapter Six – Experiments Are They Part of the Scientific Method?, Chapter Seven – Proof Is For Alcohol

Gravitational waves don't travel. Firstly, "gravitational waves" is a nonsensical term. Gravity is the phenomenon of topic, the concept of attraction between two or more objects, and as such can not wave, or travel, be moved, warped, or otherwise manipulated. In science we never move motion, or say that a concept travels. Gravity is instantaneous over any distance. The Quantum magicians would have you believe that if the sun disappeared right now, earth would maintain her orbit until the mediator of gravity "caught up with her" about 8 minutes 20 seconds later. We cover, with the luxury of detail, what the mediator of gravity IS in the book, Rope Hypothesis and Thread Theory available on Amazon.

Secondly, the speed of light c is constant through any mediator. Speed is NOT measured, angles are. Measurements of angles of light and accompanying calculations fool the particle physicist (Snell's Law). The mathemagician holds wavelength constant allowing c to vary. But wavelength and frequency are inversely proportional precisely because c does NOT ever vary and light is neither a particle nor a wave! As light "travels" through different media its wavelength and frequency changes, c remains constant.

Rational Science Vol. II; Chapter Fifteen – The Nature of Light, Chapter Sixteen – Light ...Does It Travel Rectilinearly or Curvilinearly?, Chapter Seventeen – Distance To the Stars, Chapter Eighteen – Shapiro Effect, Chapter Nineteen – Distance ...The Rubber Ruler; Rope Hypothesis and Thread Theory; Chapter Thirty – What is a Shadow?, Chapter Thirty One – Light, Gravity, and Magnetic Moment, Chapter Thirty Eight – How is Sound Different Than Light?, Chapter Thirty Nine – The World's First Flashdark

Let's sort through more of these abstractions, shall we? "Since gravitational waves don't interact strongly with other masses, there shouldn't be any dispersion as they travel through the vacuum of space. "

The concept "gravitational waves" can not interact with objects or other concepts, so there is no"thing" to be dispersed. Also, space is nothing, as such it is impossible to travel through "it."

Rational Science Vol. I; Chapter Seven – Space

"However, there is another way that dispersion might occur. If gravitons have mass, then gravitons with different energies would travel at different speeds. Over the course of 3 billion light years this dispersion would be big enough to observe. "

Of course a zero mass particle can not possibly exist, let alone be dispersed! This is the typical ad hoc theory formation to explain away failure of a hypothesis' claim of zero mass. When star gazers can not detect gravitons they reinvent them. These people never stop to challenge their hypotheses, erase the white board and start over. They just add layers of absurdities on top of irrationality.

Rational Science Vol. I; Chapter Seventeen – Mass Part One, Chapter Eighteen – Mass Part Two, Chapter Nineteen – Mass Part Three (energy according to relativity)

"The latest merger showed no evidence of dispersion, which means gravitons (assuming they exist) appear to be massless. General relativity passes yet another test."

How convenient! When QM fails, all hail Relativity. But, no, we do not assume massless zero point particles exist, or that relativity passes another test. Relativity fails conceptually at every level of reason, logic and math.

Rational Science Vol. I; Chapter Nine – E=mc Squared Away, Rational Science Vol. II; Chapter Twelve – Dimensions, Chapter Thirteen – Dimensions of Reality, Chapter Fourteen – The Three Dementia of Geometry, Chapter Twenty Six – E=mc^2

Brian finishes up his article with appeals for more detectors and more time:

"The next step for gravitational astronomy is to bring more detectors online. With the limited data we have, we can't pin down the exact location of these mergers in the sky, so we can't connect a merger event to things seen in the optical spectrum. More detectors will also let us gather more information about the rotation of black holes before their mergers, which will let us further test general relativity."

The next step for gravitational astronomy is to stop looking for gravity, Brian. Gravity is not a thing; it is what some"thing" is doing. All the detectors in the universe will never detect a non-existent gravitational wave or black hole merger. These are irrational abstractions that no amount of data can reify into a noun of reality. The mediator of gravity is a physical object far smaller than the smallest wavelength of light, so naturally no optical device can ever detect it. No amount of tests can justify relativity's irrational proposals of warped space or gravity wells. Relativity and Quantum magic will never find a common ground for gravity in Standard particle Physics except this: particles can NEVER explain attraction.

Again, what ARE gravitational waves?

"Gravity waves: They were first proposed by Henri Poincaré in 1905 and subsequently predicted in 1916 by Albert Einstein on the basis of his general theory of relativity. Gravitational waves transport energy as gravitational radiation, a form of radiant energy similar to electromagnetic radiation." - WIKI

Rational Scientific Method Vol IV; Chapter Four – Knowledge and Prediction, Rational Science Vol. I; Chapter Eight – Expanding Universe, Chapter Nine – E=mc Squared Away, Chapter Ten – Einstein

Before we continue with evidence and experimentation, let's briefly see if we can understand what the difference is between Bosons and gravitons?

"Bosons have, by definition, integer spin. The Higgs has zero, the gluon, photon, W and Z all have one, and the graviton is postulated to have two units of spin. Quarks, electrons and neutrinos are fermions, and all have a half unit of spin. This causes a huge difference in their behaviour."

Rational Scientific Method Vol IV; Chapter Twelve – Higgs Boson – What Is It? (A Parody), Chapter – Thirteen – The Higgs Fake (A three part critique of Alexander Unzicker's book), Chapter Fourteen – The Higgs Fake - Part Two, Chapter Fifteen – The Higgs Fake – Part Three

What is the difference between Higgs Boson and Gravitons?

It's spin, baby! If the difference is spin and "This causes a huge difference in their behaviour," it behooves us to understand what spin is.

What is spin?

"In quantum mechanics and particle physics, spin is an intrinsic form of angular momentum carried by elementary particles, composite particles (hadrons), and atomic nuclei. Spin is one of two types of angular momentum in quantum mechanics, the other being orbital angular momentum." - WIKI

Jon Butterworth informs us, that, "The difference between them is just spin. But in this context, spin is a quantum number of angular momentum. It is a bit like the particle is spinning, but that is really just an analogy, since point-like fundamental particles could not spin, and anyway fermions have a spin such that in a classical analogy they would have to go round twice to get back where they started. Quantum mechanics is full of semi-misleading analogies like this."

Yes, Jon, we have gotten used to misleading analogies like this. In case analogies don't float your boat let's get back to mathemagics:

"In quantum field theory (QFT) all particles are described and interact via equations on the states and operators. Integer spin particles obey commutation relations on their operators. Half integer particles obey anti commutation relations - otherwise there is no consistency or infinities in the theory that can not be renormalized. It goes deeper, and the why and how that translates into so many different properties for fermions and bosons would be a linger story. So the graviton is the PRESUMED quanta of the gravitational field."

There's that word again. If you want to read more about renormalization try this: Renormalization by John Biaz http://math.ucr.edu/home/baez/renormalization.html

But in reality, spin is either clockwise or counter clockwise. Just watch King James as he spins a basketball on his index finger. The math, whether renormalizable or not, has nothing to do with spin!

Physics stack exchange thinks it might help us to distinguish between mass and gravity. We are told that without gravity, there would still be mass. "The Higgs gives particles mass (inertia) which would exist even if there was no gravity. The graviton is the hypothetical particle that carries the gravitational interaction between massive particles."

In String theory the graviton is a boson, but in Loop Quantum Gravity, the graviton doesn't exist. "Gravitational waves can exist without gravitons." - Robert Berezdivin, PhD Physics, General Relativity/Cosmology.

So, we have the Higgs field which is separate from the gravitational field, and depending on who you are talking to, gravitons do not exist! Well, that certainly clears a lot up for me - not! Let's move on to some of the experiments proposed to determine what all this means.

Well, this is embarrassing! I said in the last chapter that we would be looking at evidence, and I even entitled this, "Evidence for Quantum Gravity." But little or no evidence has been forth coming. At least we learned that evidence is NOT part of the scientific method. What is evidence for a String Theorist is not for a Loop Quantum gravity theorist!

Chapter Nine - Quantum Experiments

We'll look at some but first let's get something straight. Experimentation is extra-scientific and therefore not part of the rational scientific method. Aristotle conceived of a spherical earth long before Eratosthenes "predicted" it with math, and Magellan "proved" it by circumnavigating the globe. We are told that the idea is to understand cause and effect using an experiment which utilizes controls and variables. But understanding does NOT come from guessing, experimentation, right, wrong, or proof. Understanding comes when one can conceptualize the objects and rationally explain the phenomena. Experiments describe. Descriptions don't explain anything. We describe objects, we explain phenomena. We point at an apple, describe it, or offer a photograph of it. We attempt to explain WHY the apple fell onto Newton's head. We don't describe it falling at so many meters per second squared. A kindergarten child understands that if the apple falls from the tree it moves real fast and lands on the ground. BUT WHY? Why doesn't the apple fall up into the sky? This is what everyone really wants to understand. Who cares how many persons can repeat an experiment if we don't actually understand anything?

Rational Science Vol. II; Chapter One – The Rational Scientific Method, Chapter Two – Scientist, Science, & The Scientific Method, Chapter Three – Scientific Method? For Dummies!, Chapter Four – Hypothesis, Theory, Conclusion, Chapter Five – Science & Technology - Conceptual & Empirical, Chapter Six – Experiments Are They Part of the Scientific Method?

What Are Some of the Experiments?

Free-fall experiment could test if gravity is a quantum force

Anil Ananthaswamy, of New Scientist gives the quantum prankster hope in this article about an experiment at the University College of London that will give answers about "whether gravity is quantum or not, to settle questions about the force's true nature." Of course, these experiments are looking to reconcile relativity with QM and hence are based on warped space, particle bias, Big

Bang creation and non-existent black holes. It all breaks down, says Anil, at "paradoxes and things like singularities."

Rational Science VOl III: Chapter Twenty Six – Big Bang Theory, Chapter Twenty Seven – Black Holes

Assuming gravity is a "quantum mechanical force" two free-falling masses, in superposition find themselves "two place at once." Being in this state could cause them to "get entangled by gravity such that measuring the properties of one mass could instantly influence the other." The experimenters will drop a "neutrally charged mass" through a continuously varying magnetic field. Based on whether its spin is up or down, the mass will take different paths. Microwave pulses will change the spin at various points along the masses path and when it reaches the bottom of its descent, "the paths then come together again and the mass is brought to its original state."

It is difficult to find two "masses two places at once," especially if they are nowhere at all! I hope they are talking about objects since they give them specific weights in kilograms, and they are dropping "them" through magnetic fields. Of course, there are different definitions for mass such as "a property of a physical body" which begs the question, how can a property be anywhere, or DO anything?

It is not clear which mass they are using. They appear to mean mass as in: "the property of matter that measures its resistance to acceleration. Roughly, the mass of an object is a measure of the number of atoms in it. The basic unit of measurement for mass is the kilogram." Since we are looking at gravity, perhaps they mean gravitational mass as in this Encyclopedia Britannica definition: "in gravity: Gravitational fields and the theory of general relativity. Gravitational mass is determined by the strength of the gravitational force experienced by the body when in the gravitational field g." But, in Quantum field theory, according to this wiki entry, "Quantum mass manifests itself as a difference between an object's quantum frequency and its wave number."

Is an object's mass its resistance to acceleration, the number of atoms it is comprised of, or is it the frequency and wave number? No wonder, they are having difficulty reconciling GR with QM.

We will never make it through the first experiment if we get into neutrally charged mass, so I'll refer you to a few chapters in a book for now.

Rational Science VOl III; Chapter Twenty Three – Mass Part One, Chapter Twenty Four – Mass Part Two, Chapter Twenty Five – Mass Part Three, Rational Science Vol. V; Chapter Thirty Three – Ions, Charge and Matter, Chapter Thirty Four – Elementary Charge

So Pinky and the Brain will drop two masses through changing magnetic fields and zap them with microwaves to alter their state, or is that their path of descent? "Each mass has two possible paths. One of these states represents paths in which the masses come closest together." Each mass will have two possible paths, or is that states, and so combined there are four possible paths, or is that states? A path is a trajectory or itinerary, a state is a condition something is in. It's not clear, but let's roll on. When the masses have returned to their original state, they will test their spin. If their spin is entangled, that could mean that gravity is a quantum force.

What do they mean by entangled? Science Daily defines quantum entanglement as, "a quantum mechanical phenomenon in which the quantum states of two or more objects have to be described with reference to each other, even though the individual objects may be spatially separated."

It is assumed that electromagnetism or Casimir force is not involved, and also that a null result does not mean a classical interpretation because one would have to rule out environmental conditions like photons or molecules colliding with the masses.

And finally this: "The biggest hurdle to carrying out the experiment for real would be putting such relatively large masses in a superposition. The most massive objects that have been observed to be in two places at once are still orders of magnitude smaller than what is required here." Interesting use of the term "real." Does this mean that the two masses are not real? Is this all done virtually in a computer program? AND what do they mean by observed? Surely no one has ever seen any object two places at once.

Avery Johnson of Popular Mechanics in March of 2018 wrote an article entitled:

Scientists Propose Experiment to Test Quantum Gravity

They propose an experiment to see if it is even possible to unify gravity with Quantum mechanics. Einstein's warped space is a given, scientists "know how it works, but not really what causes it." In other words, the sun curves the space around it pulling in the earth and moon, but how does space curve? Good question! I thought warped space was what caused gravity, but now you're asking what makes space curve? Isn't that circular?

"One possibility is that the force of gravity is carried by a particle…" This is called the graviton. Just like the photon carries the force of electromagnetism, the graviton carries the force of gravity…yeah…that's it! But the problem is that everything that has been discovered in the 20th century resists "all attempts by physicists to combine them into one grand unified theory, leading some physicists to question whether they can even be combined at all."

To make things worse, "The biggest roadblock for physicists is that the graviton, if it exists, is so weak that it's pretty much impossible to detect. In order to spot it, we would need a detector so massive it would collapse into a black hole." I wouldn't worry too much though, since, "This presents a bit of a problem that requires some extreme creativity to solve," I'm certain they'll be able to come up with something. After all, as Marge Simpson said, "We can do anything now that science has created magic."

A couple of groups of scientists have proposed experiments that can help them find out "if the graviton exists without ever observing it." How are they going to do this? Why they'll use gravity to entangle two particles. No problem. Scientists have been entangling particles for some time now. They will take two very tiny diamonds and drop them through a vacuum chamber studying their gravitational strength. "If there's a quantum element to their interaction, the strength of the gravitational force should vary."

If the experimenters succeed it would prove gravitons exist and that the marriage of QM and GR is possible. Then, all they need is a theory of how this marriage would come about. These people just don't have their feet solidly on the ground. It goes with the territory, quantum gravity is like that, don't ya' know?

Never mind that there are no viable theories for gravity using discrete particles. There never will be because one can not mediate pull with particles, no how, no way! Never mind that General relativity and Quantum mechanics are irreconcilable. Neither warped space nor massless particles are possible. Never mind that experimentation is extra scientific and that "you see what you want to see, and you hear what you want to hear." Just ask any Quantum magician to define "exist" and observe their entire make believe world as it comes crashing down around them.

This next article from physics world dot com, entitled:

"Physicists propose space mission to test quantum gravity" Written 28 May 2018, by Michael Banks is out of this world! Literally. A world wide collaboration of researchers want to use the International Space Station "to test the fundamental nature of quantum mechanics." That's all well and good, but we don't need billion dollar space stations to test the basic assumptions of Quantum mechanics. They have been weighed and they have been measured and they have been found wanting.

QUEST wishes to test "whether gravity can affect a quantum state of light over large distances by firing entangled pairs of photons from a ground station to the ISS." If the European Space Agency gives the go ahead this should happen by 2020ish.

Timothy Ralph from the University of Queensland and his associate researchers believe "that quantum states should behave differently than classical counterparts under the influence of gravity." Both GR and QM have been separately tested very extensively and passed all tests with flying colors "to incredible precision" but they are so different from each other on a fundamental level it is difficult to pound them, even with large hammers, into the same box. That will never stop these guys from trying, however.

According to the New Journal of Physics, they will send some photons to ISS, and then entangle some photons and fire them at the space station. The arrival time of photons should be different by a small percent and that will confirm, in their minds, quantum gravity.

Originally there was a proposal sent to the European Space Agency to test "quantum communication" using the space station but this secondary proposal seems more promising. Duh! It makes sense to check and see if there is such a thing as gravitons before assuming they do exist and trying to test communication systems designed to work with them! Some one here is showing some common sense for a change.

"The ground station will be used to generate the single photons using a faint pulsed source." Really? Have any of these guys actually read the literature and the specification sheets on single photon emitters and detectors?

Single Photon Emitting Diodes, a Review:

QD based SPEDs:

Single Photon detectors:

http://www.photonics.com/Article.aspx?AID=46177

"Quantum key distribution encodes binary '1' and '0' values onto photons in such a way that an eavesdropper performing a measurement on the photons will change the state of a sufficient proportion of the photon streams to leave a 'fingerprint' which can be detected by the authorized users." - Professor Gerald Buller, School of Engineering and Physical Sciences, Heriot-Watt University, Edinburgh, UK.

Read through the scholarly literature carefully.....and one sees that they are talking about photon states where all "states of light are confined to a region defined by a wave packet."

Many are produced by Quantum Dots which are nanoparticles in semiconductors that produce 0.1 photons per pulse with a 5% probability of more than one photon- (Bennett & Brassard)

The Hansbury Brown Twist Experiment uses wave packets for autocorrelation where they compare the rate of multiple photons to single photons emitted. In the article "Single Photon Emitters" by Abraham Slachter, he discusses various kinds of "single" photon emitters. Quantum Dot Single Photon Emitters (QD SPEs). These are, as he says, "structures confined to a 3-d space so small that particles in the QD start to experience quantum mechanical effects due to their confinement." This is called Quantum confinement effects. The device emits one or more single photon states. In the author's experiment using single photon wave packets for his single photon emitters, "an average wave packet from 0.7 to 0.8 could be found which is good enough for some proposed quantum computing experiments."

https://www.rug.nl/research/zernike/education/topmasternanoscience/ns190slachter.pdf

"A sequence of single photons is emitted on demand from a single three-level atom strongly coupled to a high-finesse optical cavity. The photons are generated by an adiabatically driven stimulated Raman transition between two atomic ground states, with the vacuum field of the cavity stimulating one branch of the transition, and laser pulses deterministically driving the other branch. This process is unitary and therefore intrinsically reversible, which is essential for quantum communication and networking, and the photons should be appropriate for all optical quantum information processing."

http://prl.aps.org/abstract/PRL/v89/i6/e067901

"Single Photons on Demand From A Single Molecule At Room Temperature" talks about various ways of producing "single photons" by controlled excitation of a single atom in a cavity or a single phosphorescent molecule.

"To demonstrate non-classical sub-Poissonian statistics of our source - it is convenient to compare probability distribution p(m) for our source to that expected from a Poisson distribution by means of the Mendel parameter [formula] where J2 is the variance of distribution and nav is the average number of photons."

https://www.cs.duke.edu/~thl/papers/LaBean.Nature.reprint.pdf

Key words in these articles are: wave packets, photon streams, Quantum Confinement Effects, photon states, sequence of single photons, sub-Poisson statistics, probability distribution and average number of photons.

Now, here's your homework, look through the manufacturer's specification sheets for various "single photon" emitters and detectors. You will likewise not find any actual single photon emitters or detectors, by design specifications. Finally, Nature says this, "There is still no 'ideal' on-demand single-photon emitter https://www.nature.com/articles/nphoton.2016.186

Let me translate that for you. THERE ARE NO SINGLE PHOTON EMITTERS.

Rational Science Vol. I; Chapter Twenty – Quantum Magic

Chapter Ten - Competing Quantum Theories

Quantum Gravity and the Standard Model

Before we discuss some of the competing theories, let's briefly touch on what these people are actually doing with their accelerators. And before we do that, let's recap what they are saying.

What they are saying

I found an article in the Atlantic technology archives by Garrance Franke-Ruta, who wrote it shortly after the "discovery" of the Higgs boson in 2012. It is entitled, "Still Confused about the Higgs Boson? Read This."

When it comes to sub-atomic particles there are two basic kinds, fermions and bosons. Fermions are the electrons, protons and neutrons (originally of the Rutherford planetary model). Two of these can not occupy the same place "on an atom." These are composite particles made of quarks and leptons. Two bosons, on the other hand, CAN occupy the same place on an atom. This is because, well, they aren't really like matter, they are "more like a force than a 'thing.'" None the less, in the Standard Model physicists call them force-carrying particles.

"… if this whole particle-that-lacks-mass thing is still tripping you up, you don't need to use that word in your own head; bosons lose nothing for our purposes by being thought of as entities, even if they are still technically particles, which is to say something really small of which other things are made. Some bosons have mass and some don't. The Higgs boson has a very large mass for a sub-atomic particle, though of course it is still sub-atomic, which is to say tiny. Yes, I am aware this is image looks technical and confusing. But it lays out the fundamental building blocks of particle physics and how they interact with each other. "

No, Garrance, that really does put things in perspective for us. The particle physicist uses abstract objects in their math, and ordinary words can not be used in their ordinary way when reifying these irrational terms. Also, since "I can safely say, no one understands quantum" (Feynman) any words will do. We can use spin which is not spin, color which is not color and charm or strangeness which are neither. Doesn't really matter, the Quantum mathemagician will know what we mean (nudge, nudge, wink, wink).

We have mass-less gluons which glue quarks together and carry the strong force, W and Z bosons which carry the weak force, mass-less photons which carry the electromagnetic force and now, the Quantum magicians hope to prove the hypothesized graviton which carries the gravitational force.

When we talk about the Higgs boson that was "measured" by the Large Hadron Collider, we are really talking about three things: the Higgs field, the Higgs Boson and the Higgs mechanism.

"The Higgs field is a quantum field that the Standard Model of physics predicts pervades the universe and creates drag on particles. Higgs bosons act 'as the intermediary between the Higgs field and other particles.' All fields are mediated by bosons, some of which pop into and out of existence depending on the state of the field, sort of like how rain drops emerge out of a cloud when it reaches a certain point. This interaction between the field, the boson and other particles is the Higgs Mechanism." How this mechanism works is not understood, but all bosons interact with their fields and other particles and some particles acquire mass.

After this article was written the particle guys were able to use enough energy to smash particles together with enough force to knock Higgs bosons out of the Higgs Field and actually, like really, measure them...therefore, for sure, prove their existence!

"Think of it a little like this: by smashing things hard enough, a little bit of the Higgs field got chipped off into a boson that could be measured before decaying. Sort of like throwing a rock really hard at a concrete wall -- eventually part of the wall will chip off. In this case, it was like a wall that only threw off a little bit of dust in response to a major collision, and then scientists were able to tell that the wall was there because they took a picture of the dust before it blew away. Except in this case the wall is also continuous and infinite, and invisible, and we all live inside of it, and it's what gives us mass, which is to say the quality of physical existence.

Which kind of explains why some have called the Higgs boson the God particle."

Thank you, Garrance for putting the Quantum religion into some language that the layperson can actually understand! That was less confusing than Professor Dave Kornreich, PhD, of the Department of Physics and Physical Science at Humboldt State University in California. We know we are supposed to believe him since he got his PhD from Cornell University and founded Ask the Astronomer. However, when he says, "In the quantum universe forces are mediated by virtual bosons, photons and gravitons. The key here is that these particles are virtual, not real. They exist of course, so in that strict sense they are real, and en masse can be detected and measured"…it is very confusing. How can something that is virtual, or not real, also be real and exist? Well, now that we know, we will just tell ourselves they are real because math predicts them and experiments prove them!

What They Are Doing

Let's turn now to CERN's very own website and see if they can tell us what they are doing with their accelerators. That's right! There are several of them linked together. This allows researchers to "speed up and increase the energy of a beam of particles by generating electric fields that accelerate the particles, and magnetic fields that steer and focus them."

Accelerators can be giant doughnut rings or long straight tubes. Linking these together allows higher energies. Different types of particles can be used depending on the experiment, but the Large Hadron Collider (LHC) typically smashes together protons and heavy lead ions. They get their protons from a handy can of hydrogen for making beams in the 27 kilometer ring.

Briefly, an electric field rips the single electron away from the hydrogen nucleus and starts the proton on its journey along the tube as alternating electric fields at specific frequencies pull the charged particle along. By manipulating the frequency along the ring the CERN engineers "ensure the particles accelerate not in a continuous stream, but in closely spaced 'bunches.'"

Radio waves interact with these bunches of particles as they pass by special metallic chambers. The "electric field in an RF cavity" transfers some of its energy to the particles "nudging them forwards." Wait! Maybe I'm being nitpicky, but is "nudging along" the same thing as "pull?"

Dipole magnets around the ring bend the "path of the beam of particles" otherwise they would go in a straight line. Quadrapole magnets focus the beam pushing the particles into a tighter beam. In this way, particle beams can be made to hit a target or two beams can be smashed together.

Particle detectors are placed around the collision site to "reveal the particles that emerge from the collisions."

Detectors detect traces of the particles in the aftermath of the collisions to identify them much like a hunter identifies an animal by its tracks. In this way they learn about their mass, charge and speed.

Detectors and layers of sub-detectors allow engineers to calculate a particles momentum from the curvature of its path. Lots of momentum means a straighter path whereas low momentum means a spiraled path. Sub-detectors may look for specific properties and calorimeters absorb and measure the energy of a particle.

"Tracking devices reveal the paths of electrically charged particles as they pass through and interact with suitable substances. Most tracking devices do not make particle tracks directly visible, but record tiny electrical signals that particles trigger as they move through the device. A computer program then reconstructs the recorded patterns of tracks."

After trackers and caloriometers do their thing, researchers have ways of detecting radiation emitted by the particles. They look at something called Cherenkov radiation given off when a particle goes through various materials. By measuring angles, which are velocity dependent, particle velocity can be calculated. If one knows the velocity and the momentum of a particle, its mass can be calculated as well. Now, based on that data, we know what the particle is. We just look at our handy dandy chart of particles and identify that rascal, forthwith!

Particle types can also be "distinguished" by looking at something called transition radiation which happens when a fast moving charged particle "crosses the boundary between two electrical insulators with different resistances to electric currents."

All of this together gives the particle guy a snapshot of a movie of the collision! Exciting huh? In a later chapter we'll take a look at the technical and methodological problems inherent in particle physics' accelerators. Here is a forward look at some of that.

With the half life of 10^{-25} seconds for a top quark, a particle can't travel more than the radius of a proton in that time, and consequently never make it out of a collision's 600 million protons in order to reach a detector. With 100 billion pairs of photons how does the particle physicist extract one pair from the decay of a Higgs boson? Yet they claim this is exactly what they have "observed." They have not observed anything, as important as that is to them. They have merely used event generators built on theoretical calculations to simulate collisions, and compare these layers of theoretical assumptions to signals received at the detectors. Who knows what is buried in the millions of lines of computer code, but it is easy to see how one can tweak the numbers to agree with the outcome. The detector signal represents energy, charge and path of the particle. Yet muons, taus, neutrinos, quarks and bosons can all be used to describe the results hoped for. There can be no objective interpretation of this experiment at all.

I Forgot - Why Are we doing This, Again?

We are trying to unite Relativity with Quantum; The Big with the Small.

"The non-renormalizability of Einstein's gravity is still an unsolved problem requiring a new theory: string theory is one of many as-yet incomplete pretenders to that throne. Whether "curvature of spacetime" is the last word or not remains to be seen: if gravity is to be quantised, then smooth, differentiability concepts like "curvature" are going to need a quantum reboot." - Andy Buckley,

PhD in particle physics, visiting researcher at CERN, lecturer in physics

Is gravity the exchange of gravitons, or the curvature of spacetime?

"Theorists believe that both of these descriptions are valid, in much the same way that we can think of the force of electromagnetism as being either the product of a continuous field, or the exchange of numerous force-carrying particles called photons. For certain 'classical' calculations, the description of electromagnetism as a field is more workable than its 'quantum' description, and vice versa. The problem is that, although physicists have a workable theory of gravity that involves the gravitational field, and gravitational forces as a curvature of spacetime, there is no currently believable quantum theory of gravity involving 'gravitons'. We do not even know, a priori, whether there are such things as gravitons even though we have isolated the particles responsible for the other three forces in nature. Just as photons are 'packets' of the electromagnetic field, gravitons would be considered 'packets' of the gravitational field or space-time curvature." - Dr. Sten Odenwald (part of the NASA Education and Public Outreach program).

Ask CERN "Why is gravity so weak and what are some of the explanations for Quantum Gravity?" This is the response you get: Extra dimensions, gravitons, and tiny black holes. In case you think I am making that up let me quote from CERN's website:

"Though it may sound like science fiction, if extra dimensions exist, they could explain why the universe is expanding faster than expected, and why gravity is weaker than the other forces of nature."

AND how do they propose to test for these extra dimensions? Well, silly, of course, "One option would be to find evidence of particles that can exist only if extra dimensions are real." Our standard particles could actually be heavier in other dimensions than they are here in our dimension. Only the LHC has the ability to create enough energy to knock these heavier versions (called Kaluza-Klein states) out of the other dimension(s) and detect them here in our dimension.

Another way of revealing extra dimensions would be through the production of "Microscopic black holes."

"What exactly we would detect would depend on the number of extra dimensions, the mass of the black hole, the size of the dimensions and the energy at which the black hole occurs. Finding more on any of these subjects would open the door to yet unknown possibilities." https://home.cern/about/physics/extra-dimensions-gravitons-and-tiny-black-holes

There are only three dimensions. Dimensions require orthogonality. All dimensions of an object must face outward at 90 degrees to each other. The three dimensions of reality (L, W, & H) do this. Dimension: any of three mutually perpendicular directions in which an object faces or points simultaneously. Rational Science Vol. II; Chapter Twelve – Dimensions, Chapter Thirteen – Dimensions of Reality, Chapter Fourteen – The Three Dementia of Geometry

What Are Problems with Quantizing Gravity?

Let's move away from the obviously optimistic CERN folks and see what others are saying. Let's look at a slightly more detailed analysis of the problems facing the particle physicist. Let's look at an Op-Ed piece at Space dot com by astrophysicist from the Ohio State University and chief scientist from COSI Science Center, Paul Sutter. The piece is entitled:

Why Can't Quantum Mechanics Explain Gravity?

It is accompanied by a NASA/ESA image simulating gravitational waves generated by merging black holes, and begins with an example problem: What happens when an electron meets a photon?

The answer is a bit complicated, because, we are told, "photons can be created or destroyed at will." After all, one creates a gazillion of these when they flick on the light switch and they are instantly destroyed when they get absorbed by atoms as they hit the wall.

"...since matter and energy are two sides of the same special-relativistic coin — just as space and time are related under a single framework — if energy can be created or destroyed, then so can mass. You can create photons, and you can create electrons. Have an empty box? Poof, like magic, they can appear! Or disappear! Whatever! Of course, if they do appear, they'll disappear right away — stealing energy from the vacuum to exist for just a little bit; that's all they can muster. But if you already have something — say, a photon zipping about — it can turn itself into an electron and positron without even realizing it. And those particles can turn themselves back into a photon if they feel like it. It's seriously just that easy. Matter and energy are really the same thing, and can change forms as easily as you change your shirt. Of course, there are rules and limitations, but a true magician never reveals his secrets. "

It seems the classical view of physics with its smooth waves and fields is old hat as far as quantum is concerned with its "jaggedy and blocky" energy levels and angular momentum. The space-time permeating electromagnetic field can "wave around" and sometimes part of it (a photon) gets "pinched off." By removing or adding energy from a particular region of the EM field one can create or destroy photons. That is classical physics. It's also creationism without God. Can someone please tell these guys that when a magician pulls a rabbit out of his hat, he gets the rabbit from somewhere.

Just as there are photon fields, there are electron fields, so if you add energy to the electron field... out pops an electron! Now trying to answer the original question about what happens when an electron and a photon collide is more difficult to ascertain because, with fields, there are an infinite number of possible collisions. Since either of these gazillion photons and electrons (there are also positrons) can appear and disappear or even swap identities (that's right, one can shape shift into the other) it is impossible to add up all the possible collisions.

"When infinities crop up, you can't make progress, you can't make predictions, you can't make science. "

Wfew! It's good to know the mathemagicians do have their limits! Not to worry, though, good old Dick Feynman and others came up with a few "tricks," some mathemagical terms that allows them to replace infinite numbers of places in equations with just a "handful of places." That made quantum workable. Just plug in some known values like electron mass, and even though we can't predict everything, "it has made incredible progress in describing the interactions of all particles and three out of the four forces."

Now, what about gravity?

Gravity isn't a force like the other forces. Gravity is related to space-time, the "stage on which all the particles" are "the actors." Not only do the actors move around on the stage, the stage moves around as well. "It bends and warps under the influence of the actors, and that bending and warping redirects the actors' motions."

We will look at Einstein's field Equation in greater depth in a future article on mathemagical incantations and spells, but for now suffice it to say that:

The left side of the equal sign is "Spacetime tells matter to move" and the right side of the equal sign says, "matter tells spacetime to curve."

$$G_{\alpha\beta} = \frac{8\pi G}{c^4} T_{\alpha\beta}$$

To make things more complicated, besides having to account for every possible combo of electron and photon, one must consider "all possible permutation of spacetime." This is far too complicated and the mathematical models break down.

String Theory and Loop Quantum gravity are nice ideas but neither have testable predictions. This is why black holes and Dark energy are so appealing. It has to do with something called the cosmological constant and expanding space.

According to Brian Greene on his PBS series and in his book entitled The Fabric of Space, "Energy forms the fabric of space. We call this energy Dark Energy." In the chapter entitled String Theory he writes, "Gravity pulls chemical elements and dark matter into stars and galaxies and repels dark energy into a thin quantum field less than 4 electrons volts per cubic millimeter."

Greene tells us that gravity has a repulsive force which accounts for the universe expanding at an accelerating rate - that space itself is expanding as well, and that there were multiple big bangs.

This relates to the cosmological constant which is the value of the energy density of space. Einstein gave this a value in his equations until it was "discovered" that space was expanding. He called this his greatest blunder and changed the value to zero.

If there is a positive value to this number it means that the universe is expanding at an accelerated rate and if negative then plain old ordinary matter would decelerate universe expansion.

Vacuum energy density is part of the Higgs mechanism for something called spontaneous symmetry breaking. Just after the Big Bang creation of matter, space and time, there was a lot of this "stuff" and it must be related to "negative pressure" or what Greene referred to as gravity's "repulsive force."

If the universe grows (mass creation) and vacuum energy density stays the same, then matter energy density would decrease and gravitational acceleration would also decrease; if the universe got smaller, then the gravitational acceleration would increase at an accelerated rate. Each minute change would be magnified and so the universe would be unstable and the model of Einstein is flawed. Therefore the universe can not be static. Hubble's constant showed the static universe was incorrect and so Einstein removed the cosmo constant from General Relativity. Doesn't matter to these guys that matter creation isn't possible, only that their math "predicts."

There is a "surprising consequence of a positive cosmological constant," says David H. Bailey in his 2017 article by that title: "The paradox derives from the fact that when one calculates, based on known principles of quantum mechanics, the "zero-point mass density'" or the "vacuum energy density" of the universe, focusing for the time being on the electromagnetic force, one obtains the incredible result that empty space "weighs" 10^{93} grams per cc. Since the actual average mass density of the universe is 10^{-28} grams per cc [Susskind 2005, pg. 70-78]." Yes, that's right, doesn't matter to these folks that space is nothing, void, zip! Space is magnitudes stiffer than steel because that is what is predicted by their math.

Since this is 120 orders of magnitude greater than what is measured, it presented quite a "paradox." Supersymmetry didn't pan out for the Quantum magicians, but they still remained hopeful that they would come up with some math that would allow the mass energy density and vacuum energy density to cancel out to zero.

Unfortunately, Hubble determined that the universe was expanding at an accelerated rate and that dashed all hopes of a zero point mass density. To make things worse, a few years ago, NASA/ESA used the Hubble telescope to measure distances to stars in nineteen galaxies with a higher precision and discovered "the U" is expanding faster than indicated by measurements they took "just shortly after the Big Bang."

Rational Science Vol. I; Chapter Eight – Expanding Universe, Rational Science Vol. II; Chapter Thirty Five – Dark Energy and Dark Matter

The Hubble folks announced as recently as 2018 that they measured cephieds at 10 times further away than previously and so the universe is expanding even "faster than we thought."

Rational Science Vol. II; Chapter Seventeen – Distance To the Stars, Chapter Eighteen – Shapiro Effect, Chapter Nineteen – Distance ...The Rubber Ruler

Dark Energy is the force that is accelerating the universe and accounts for the 70 percent missing mass. This is tied into the idea that the universe is nearly flat. Steven Weinberg proposed that the cosmo constant "must be zero to within one part in roughly 10^{120}, or else the universe either would have dispersed too fast for stars and galaxies to have formed, or else would have recollapsed upon itself long ago [Susskind2005, pg. 80-82; Weinberg1989]."

To solve the cosmological paradox, String theorists propose the existence of other universes on the order of ten to the 500th power!

What Are Some of the Theories of Quantum Gravity?

String theory, Loop Quantum Gravity, Euclidean Quantum Gravity, Causal Dynamical Triangulations, spin foam models, causal set theory, shape dynamics, twistor theory, and asymptotic safety.

The attempt to unify small scale and large scale phenomena will continue for a very long time, as Lawrence Kraus said in this interview. http://nautil.us/blog/lawrence-krauss-versus-freeman-dyson-on-gravitons:

"Now BICEP may not have seen gravitational waves from the early universe, but the fact that we recognize that if this phenomena called inflation happens in the very early universe, and if it produces gravitational waves, that will tell us that gravity is a quantum theory; therefore, all of the problems of quantum gravity will need to be addressed by theorists, giving job security for generations." Uh huh!

Quantum Gravity

In an attempt to unite the large scale with the small scale phenomena using quantum gravity, folks at the Perimeter Institute are trying to come up with experiments to test many theories: loop quantum gravity, spin foam models, string theory, causal set theory, shape dynamics, twistor theory, and asymptotic safety.

They have some ideas that they recently came up with; something called relative locality, dealing with spacetime, and shape dynamics, dealing with a "reformulation of General Relativity" and "consistent boundary formation" to redefine Quantum gravity theories. PI researchers will still continue working with quantum loop theory, casual dynamical triangulations, causal sets, and string theory. So, there are theories a plenty, and yes, they will make the foot fit into the shoe, no matter what! https://www.perimeterinstitute.ca/research/research-areas/quantum-gravity

So What Are Some of the Problems Plaguing Quantum Gravity?

Besides needing impossibly large, planet sized detectors, there is the interpretation of "the wave function of the universe" - the time problem. What is the time problem with Quantum Gravity?

With QM, time is absolute, but with relativity, time is relative. QM's time is steady, but relativity's time dynamic and "interwoven with directions x, y and z into a four-dimensional "space-time" fabric. The fabric warps under the weight of matter, causing nearby stuff to fall toward it (this is gravity), and slowing the passage of time relative to clocks far away. Or hop in a rocket and use fuel rather than gravity to accelerate through space, and time dilates; you age less than someone who stayed at home."

If that isn't ridiculous enough for you, this aught to make you belly laugh:

"Erik Verlinde of the University of Amsterdam argues that dark matter is an illusion caused by the holographic emergence of space-time from quantum entanglement."

"Researchers have worked out the math showing how the hologram arises in toy universes that possess a fisheye space-time geometry known as 'anti-de Sitter' (AdS) space."

"Geometric patterns, such as the amplituhedron that describe the outcomes of particle interactions also suggest that reality emerges from something timeless and purely mathematical. It's still unclear, however, just how the amplituhedron and holography relate to each other."

https://www.quantamagazine.org/quantum-gravitys-time-problem-20161201/

Researchers are so confused about what is real and what is mathemagic they have lost all touch with reality. Time and space are human concepts that Father Universe has nothing to do with. A concept is a relation between objects. Time is the relation between locations of objects and requires the memory of an observer. Since science deals with objects, and the rational scientific method removes the observer with his biases and limited sensory system, time is NOT a scientific term. As pointed out to me in yet another conversation on time. Time is not a scientific term. It is not a part of science because we cannot explain anything with time. Time is 100% descriptive. Time cannot tell us WHY something happened, what caused the phenomenon.

See The Rational Scientist, April issue, page 54 "What is Time?" Rational Science VOl III; Chapter Twenty – Time Part One, Chapter Twenty One – Time Part Two, Chapter Twenty Two – Time Part Three

Chapter Eleven - Loop Quantum Gravity

Let's take a more detailed look at one of the biggest "contenders" for Quantum Gravity, Loop Quantum Gravity (LQG). We will try to follow along with Lee Smolin, author, theoretical physicist, founding member of the Perimeter Institute for Theoretical Physics, and "co-inventor" of Loop Quantum Gravity, as he describes to us the challenges. We will pull mostly from Lee's Edge article here: https://www.edge.org/conversation/lee_smolin-loop-quantum-gravity-lee-smolin

Keep in mind this is an old article from 2003. We just want to see if what was predicted then has panned out. Lee talks about what he calls, "The Two Best Developed Approaches,"

Loop Quantum Gravity and String Theory (ST)

He sees that there are lessons to be learned about space and time and "how science works." I can agree before even reading another word, but I doubt we both think this for the same reason. Let's see. Based on his first few paragraphs about scientific democracy, I'd say he has an entirely different view of how "science works," Not only on the method of inquiry, but on the Peer review process, advancing in one's career, or getting published. See: Rational Science VOl II Chapter Forty Four – Peer Review

Smolin mentions relativity and quantum and the "problem of unifying them." He sees this as, "the main open problem in physics left for us to solve in this century." It is not. The main problem in physics is in NOT defining their most crucial Key Terms: object and exist.

I agree with him when he says "Nature is a unity," but trying to unify, two irrational and irreconcilable "theories" is patently ridiculous. What is needed is to do away with the current scientific method, relativity, quantum and in fact all of particle physics. The unity of nature is found in the fact that there are no discrete particles. No fields, no space and no time as "physics" currently see them. Physics is in quotes because in the entire particle zoo, the magician has NO physically present objects, only abstract concepts.

In a future chapter we will cover a rational hypothesis and explain how nature actually may perform her gravity without the use of waves or fields or any of the other quantum magic or spells and incantations of mathemagics. We will use just simple, physical, illustratable mechanisms. We will also unify gravity (pull) with the only other force – push as observed through light and electromagnetism.

Lee states the obvious when he says "the pen falls" but can point to no real physical mechanism for why the pen falls, or show any progress in unifying relativity with quantum using gravity because gravity can never be explained with discrete "particle's" push, negative momentum or warped space. If the "invention" of the two approaches (LQG and ST) was a turning point, it was a bad move. They entered into a cul-de-sac where they will turn around and around in circles forever with only one way out. Turn around and go back! Go back to before loop quantum gravity and string theory, go back to before general relativity and the Standard Model Particle Zoo, and start over. It is no wonder their calculations are "predicting surprising new phenomena" theoretical mathematical physics is getting more and more abstract as they go around the cul-de-sac again and again. They have gotten dizzy and need to stop for a while so their minds can catch up to reality. Their experiments to test their predictions will never get them anywhere, but around the circle again and to the next pay check.

Let's quote Lee Smolin as he summarizes their current predicament:

"In quantum theory, distance is inverse to energy, because you need particles of very high energy to probe very short distances. The inverse of the Planck energy is the Planck length. It is where the classical picture of space as smooth and continuous is predicted by our theories to break down, and it is some twenty powers of ten smaller than an atomic nucleus. Because the Planck scale is so remote from experiment, people began to put great trust in mathematics and theory. There were even some string theorists who said silly things like 'From Galileo to 1984 was the period of modern physics, where we checked our theories experimentally. Since then, we work in the age of postmodern physics, in which mathematical consistency suffices to demonstrate the correctness of our theories and experiment is neither possible nor necessary.' I'm not exaggerating; people really said things like this."

Sorry to have to tell you this Lee, but neither math nor experiment are part of a rational scientific method of inquiry. Particle physicists are only fooling themselves when they predict with math and confirm with experimentation. This is merely confirmation bias, because the extra scientific method of experimentation can never justify pull with discrete particles. Math is only ever descriptive and NEVER EVER explanatory.

Lee speaks of Planck scale physics and using the "Universe itself" as an "experimental device." Cosmic accelerators produce higher energy particles than can be produced here on earth in man made accelerators and send those particles to us "through the radiation and matter that fill the universe." All that is being done is changing from gathering data "here to gathering data from 'there.'" Data that can be interpreted only within the framework of the absurd proposals of GR and QM.

Planck scales can not be blown up to astronomical scales via an inflated and expanding universe. Only person's egos will be inflated and budgets will be expanded. The term "Planck scale" mostly refers to magnitudes of energy, space and time. These are three mathematical abstractions which have no physical presence. All three are concepts, and one can not use concepts as if they were objects.

Hopefully, we can get on to Loop Quantum Gravity and String Theory now, Lee.

The "basis of quantum" has been the Wheeler-DeWitt equation, which remained unsolved for quite some time but Mr. Smolin and his cohort have now "solved it exactly, in fact we found an infinite number of exact solutions." This is the sort of nonsense that pervades quantum. These people haven't got their feet on solid ground. Their heads are in the clouds. The definition of infinite, according to Oxford dictionary, is: - limitless or endless in space, extent, or size; impossible to measure or calculate. Exact means: not approximated in any way; precise. Solution is a means of solving a problem. An "infinite number of exact solutions" is the same thing as saying NO SOLUTION AT ALL!

He continues, "This has led to a detailed theory that gives us a new picture of the nature of space and time as they appear when probed at the Planck scale." I have news for you, Lee. You can't probe the nature of space or time at ANY scale. None-the-less we are told at that scale, space is made up of discrete particles. The minimum volume of the smallest unit of space (void) is the Planck length cubed, or 10^{-33}cm. Oh it gets even better:

"A surface dividing one region of space from another has an area that comes in discrete units, the smallest of which is roughly the Planck length squared. Thus, if you take a volume of space and measure it to very fine precision, you discover that the volume can't be just anything. It has to fall into some discrete series of numbers, just like the energy of an electron in an atom. And just as in the case of the energy levels of atoms, we can calculate the discrete areas and volumes from the theory."

If you corner Lee, I'm not sure which definition he will give you for surface, but the dictionary has this typically:

A) The outside part or uppermost layer of something (often used when describing its texture, form, or extent). "The earth's surface"
B) A continuous set of points that has length and breadth but no thickness.

If A) then area and region with a surface makes no sense. There can be no surface without an object. A region or area is NOT an object. Space is not an object; it is the lack of shape, nothing, zip, nada.

Likely he is talking more along the lines of B), because they are making calculations using Planck length and trying to combine QM with relativity which is based on geometry. In geometry a surface is a concept like a 2 dimensional plane with length and width but no height. Just like the Planck length. It has only length as it is typically defined from c, G and barH. This magnitude is really distance traveled, but that is a discussion for another time.

Anyways, the claim is that there have been detectable effects of these particles coming from particle accelerators made by galaxies far, far away in the form of (Now here comes the switch; particles to waves.) cosmic rays and gamma ray bursts.

"These effects are caused by light scattering off the discrete structure of the quantum geometry, analogous to diffraction and refraction from light scattering off the molecules of the air or liquid it passes through."

No, Mr. Smolin, the concept "waves" can not scatter off of "the structure of quantum geometry" if that structure is like molecules. Why do I say this? Because all phenomena are the result of surface to surface contact between two or more existing objects. Adding nonsense to irrationality, many of these cosmic bursts are "possibly caused by mergers of …black holes. If cosmic rays are light, then these impossible objects, black holes, by definition, can not emit them because light can not escape from a black hole. "A black hole is a region of spacetime exhibiting such strong gravitational effects that nothing—not even particles and electromagnetic radiation such as light—can escape from inside it." – wiki

The jist of it is that over very large distance of these rays traveling from galaxies to us, the effects of quantum gravity are magnified "to the point where they can be observed." And since elementary particles travel like waves, maybe the same effects are responsible for what they observe with the cosmic rays.

Enter the Conflict

Some of Loop Quantum Gravity's predicted effects are in conflict with the Special theory of relativity's (SR) speed of light [c] which is supposed to be constant and independent of an observer. No problem, just quantize SR and modify LQG.

Quantum theory predicts that the speed of light varies with the energy of photons. Two photons from billions of years ago should arrive at earth at slightly different times. When two gamma ray photons of different energies arrive at the GLAST observatory, the difference should be detectable. "We very much look forward to the announcement of the results, as they will be testing a prediction of a quantum theory of gravity."

Well I can't find anything that confirms GLAST results for quantum gravity on Edge, at the Perimeter Institute or anywhere else on the interwebs, but I do see lots of new experiments being proposed like this 2018 article dealing with entanglement and the International Space Station: "Physicists propose space mission to test quantum gravity." There are some others using different angles as we discussed in the previous article on experiments.

BTW: GLAST did detect a delay between gamma ray burst GW170817 and the LIGO detected gravitational waves from the merger of two black holes. This presents some problems for current theories, not the least of which is that gravitational waves travel through dense matter in collapsing stars but gamma rays are scattered by it. Also, "isolated black holes" don't create gamma ray bursts. There has to be a great deal of matter nearby for this to happen.

Loop Quantum Gravity answers some questions that the Phiz Whiz shouldn't even be asking! For instance, "If a black hole has entropy, what is the information that the entropy of a black hole counts?" QLG answers the question by quantizing the area of the black hole's horizon. Yep, "just as space is, it is made up of discrete units." But if that doesn't float your boat, Quantum also quantizes spacetime; they call this a spin foam.

Lee explains String Theory to us. Gravity is united with the other forces because "all particles and forces arise from the vibrations of extended objects. These include one-dimensional objects (hence the name "strings"), but there are also higher-dimensional extended objects that go by the name of "branes" (for generalizations of membranes)."

Of course there are only three dimensions and that does not include space or time. It includes length, width and height.

See the Best of Rational Science; Chapter Twenty – Dimensions, Chapter Twenty One – Dimensions of Reality, Chapter Twenty Two – The Three Dementia of Geometry

String Theory, we are told, although not a complete QUANTUM theory for gravity, does work for "very limited kinds of black holes" and unifies gravity with the other forces. In order for it to work we must assume there are "six or seven unobservable dimensions of space." String Theory also proposes that there are some new kinds of supersymmetry tying together "…matter like quarks and electrons with the quanta of forces like photons and gluons." Of course, we don't, but if we could see this, there would be supersymmetric particles with identical mass and charge but with a spin of one half.

We find that symmetry is broken. In other words forces are symmetric, "but the state of the world does not obey it. For example, looking around your living room, you see that the fact that space has three dimensional symmetry is broken by the effects of the gravitational field, which points down."

Yeah, I'm feeling you, Lee, Einstein's warped space has that problem too. Our sun weighs the space tarp DOWN and earth rolls merrily around it, but Charon doesn't follow this convention as Pluto weighs space outward compared to us. In other words, there is no down in space. They hope to find experimental evidence to support all these "theories" using accelerators but meanwhile, "The interesting—and unfortunate—upshot of this is that in the absence of experimental check, different communities of people have focused on different questions and invented different imaginary worlds."

Perhaps this is a shot across the bow of String theorists as Lee goes on to say he and his LQG friends continue to work on Loop Quantum Gravity in the real world where space has three dimensions. This is laughable since space is NOTHING, and as such, can have no attributes at all. I can enlarge my space closet, but I am not enlarging space, I am only increasing the distance between the walls. It's so funny to hear folks arguing over the attributes of non-existent abstract objects. Space is a where, not a what: a convenient way of organizing objects by the distance (separation) between them. Specifically, we can only conceive of an object by delineating what is inside its border from what is outside it. Anyways, as Lee says, if it turns out there are "higher dimensions and supersymmetry, they can be incorporated in to Loop Quantum Gravity." Yes, of course they can. This is what happens when one builds the ridiculous on top of the irrational ad hoc.

We discover what Lee really means about imaginary worlds when he tells us this is what happens when "science gets decoupled from experiment." See The Best of Rational Science; Chapter Six – Experiments Are They Part of the Scientific Method? No! They are NOT. Experiments are extra scientific and only confirm or deny opinions, or what you believe, not what Mother Nature is actually doing in her shop.

Abstract objects evolved from mathematics and geometry, where descriptions, not explanations ruled the day. Extra dimensions were inevitable once the theoretical physicist forgot there is a difference between objects that can and can not exist. Relativity and quantum both derive from the geometer's dementia, where a point, a line and a plane can exist all by themselves independent of any three dimensional object. Hence there are massless point particles and one dimensional strings. These are not even approximations of real objects. All objects must have shape, and the only objects that can exist, that is be located in respect to all other objects, are ones that have length, width and height.

The main crux of the argument beyond supersymmetry and extra dimensions is what is known as background dependent or background independent approaches. A background dependent approach, for example, is one where strings move "in a fixed classical spacetime" Relativity says there is no fixed background, or spacetime.

The Greek Sophists and Newton argued this absoluteness of space and time. Democritus and Einstein were "relationists." And the argument continues to this day. Mr. Smolin is very clear that proponents of LQG believe nature will always be proven to adhere to relativity's notion of a dynamic spacetime geometry. However, he sees String theorists having mostly "come around" to the idea of background independence. All the different versions of string theory have their own different fixed background. So, now most of these theorists see that there could be an infinite number of theories which are really just approximations of one grand unified theory. This is M-theory. In other words, each fixed background theory approximates a single background independent theory.

Lee basically disagrees with many of the theorists like Leonard Susskind and David Gross because of differing methodologies. They believe in working together as a community within a certain framework, and Lee Smolin believes that independent research and conflict is a more direct approach to solving the problems. I agree with lee that it is a problem of methodology, BUT that both parties need to abandon entirely their "scientific" methodologies.

I also agree that there should be "conflict and pluralism in science," but this can never happen in today's scientific environment- with Peer Review and Nobel Prize, and the entrenched "Old Guard" in our schools and institutions of higher learning. When the old Guard dies off, only then can truly new ideas be heard. Eliezer Yudowsky said, "Science moves forward by slaying its heroes, as Newton fell to Einstein. Every young physicist dreams of being the new champion that future physicists will dream of dethroning."

"Science alone of all the subjects contains within itself the lesson of the danger of belief in the infallibility of the greatest teachers in the preceding generation ... Learn from science that you must doubt the experts. As a matter of fact, I can also define science another way: Science is the belief in the ignorance of experts." - Richard Feynman

BUT, I will take this further and say that we need to abandon the empirical, mathematical description based approach entirely, because it is not the persons but the method that is failing science.

In spite of their differences in methodology and in foundational premises, Smolin tells us he had spent a couple of years working on String Theory because "I had an idea that maybe string theory and loop quantum gravity were different sides of the same theory." He had to put his work on octonions aside because there were new experimental developments with Einstein's cosmological constant that made going to work on "an imaginary world of six or seven extra dimensions" "a lot harder."

The cosmological constant has to do with the energy density of "empty space" and the cosmological data supports this, not the negative constant of String theory. String theory is incompatible with a world of a positive cosmological constant. It never occurs to either camp that "energy density" is an irrational term or that space (separation, distance, lacking shape, nothing) can "have" anything.

Since Lee sees this positive signed cosmological constant as a basis for unifying Relativity with Quantum, he tells us, "I've mostly been studying how to make predictions about the new experiments from a version of loop quantum gravity that incorporates a positive cosmological constant."

Lee ends the article with something that I wholeheartedly agree with:

"You can live for a few years in an imaginary world, but in the end the task of science is to explain what we observe."

Chapter Twelve - Crossing the Quantum Divide

A recent Scientific American magazine article entitled, "Crossing the Quantum Divide" covers another aspect of Quantum; The Quantum magicians want to bridge the gap between the macro and the micro worlds. This poses a much bigger problem than they suspect. We can only ever imagine the micro world of the atomic and sub atomic, because we can never see it. The atom is the mediator of light but the smallest wavelength of visible light is thousands of times larger than the atom. One can never see the atom. We learned, from Rational Science Vol. II, Chapter Twenty – Atoms, that atoms can't be seen or imaged by any optical system.

Theoretical physicists are using math to describe how the atoms behave. They are not actually describing how atoms look. Whether Thomson's Plum Pudding model, Rutherford's Planetary Bead, Bohr's Planetary Bead, Sommerfield's Wavon, Debroglie's Ribbon, Schrödinger's Wave Model, Born's Electron Cloud or Lewis' Shell, theorists are describing behavior not physical appearances.

Often we read that scientists have observed this or that atom using the Atomic Force Microscope (AFM) or the Scanning Tunneling Microscope (STM). The AFM scans the surface of an object with a needle tip. This device is basically "feeling" the surface or in non-contact models, detecting electric potentials. The STM uses QM tunneling currents to detect electron densities. The STM resolution is around 0.1 nm lateral resolution and 0.01 nm depth resolution. So, we are "looking" at "surfaces." Actually, we are looking at computer generated images derived from data gathered on detected electron potentials and densities. The quantum folks haven't got a clue what atoms actually look like. Furthermore, they use different models depending on which phenomenon they are trying to describe.

Imaginations rule supreme in Quantum Magic. The Quantum theorist wants to know, why is it that "the probabilistic nature of quantum" governs the "microscopic" world and yet classical physics seems to govern the "macroscopic" world? What I want to know is why use the term microscopic that way? Is it because they want to give you the impression they actually DO see something as if by a microscope?

To find out more about this Quantum Divide we travel to a lab in the Netherlands and find ourselves talking to a mad scientist by the name of Simon Groblacher. He tells us of an experiment to make a really BIG device, the size of a single bacterium. Simon wants to answer the question, "Can something be two places at once?"

That's right! QM proposes an object can be two places at once and that two objects can be at the same location at one and the same time; that Schrodinger's cat can be both alive and dead and that there are infinite possibilities, or multiple worlds where everything is happening probabilistically, until a lab rat somewhere observes it and a waveform collapses from that probability amplitude into the macroscopic world where we live and breathe and have our being. Quantum probabilities become classical realities.

Photons, we are told, are normally doing this all the time. This begs the question, "How can something be two places at once when it is nowhere at all?"

Quantum magicians propose to use the principle of superposition to make Quantum computers. Contrary to a classical bit that can only be in the state corresponding to 0 or the state corresponding to 1, a qubit may be in a superposition of both states.

My computer hard drive is a physically present object whose bits are "flux transitions" on a magnetic medium. In other words, tiny needle shaped objects embedded on a spinning plate are formatted so that they are magnetically polarized like little magnets. One end represents the 0 and one end represents the 1 in binary computing. A transition is when the poles of the needle swap. What was once 0 is now 1 and what was once 1 is now 0. How can a tiny magnet have other than the two polarities? What possible transition can one end of a tiny magnet make but it's opposite? Why, plus minus, of course! Well, instead of tiny needles the quantum computer is proposed to use photons, electrons, atoms or ions instead. We'll cover Quantum computing in more detail in a later chapter.

Simon hopes his contraption, some day, will vibrate a membrane into superposition and then the tardegrades he places on the membrane will also be vibrated into superposition. It will be two places at once! AND how can little water bear be on the membrane and yet somewhere else? Where, pray tell?

In quantum, particles do not have definite locations, or energy levels or "any other definite properties." An electron orbits its proton in a probability distribution, a region around the nucleus like unto a cloud. The electron orbital doesn't have any particular energy state, but all energy states at once. Until, of course, you look at it or measure it. This is known, lovingly, as the "measurement problem."

In spite of the fact, we are told that without quantum we wouldn't have the "electronics industry, cell phones or Google," experimenters have no idea why quantum behaves the way it does. Where does the quantum world end and the classical word begin? While theorists test such theories as decoherence and Continuous Spontaneous Localization (CSL), Simon will fire lasers at his massive millimeter sized trampoline to see if he can push it into quantum superposition where it vibrates at two amplitudes simultaneously.

No doubt, if he is successful, theorists will argue that it was two waveforms collapsing or becoming entangled or some other quantum madness. Quantum describes probabilities of where something might be, how fast it is going and at what energy level or other state it is in. Once Simon measures the membrane, the waveforms will collapse and tardegrades will be right where he put it, slightly nauseated from vibrating so fast, but right where he was before the bouncy trampoline ride. Mr. Groblacher will claim success. Other's interpretations, of course, will vary from pilot waves to many worlds, and many more experiments will need to be done before one candidate is chosen over the other. I predict that all theories will be "right." In the world of infinite probabilities, this can "happen," no?

Sorry, Simon, no! While quantum might work based on probabilities, reality works on possibilities. Something occurs only if it is possible to occur. Something exists if is physically present, not probabilistically physically present. It is here in relation to everything else that exists and not also there at the same time.

LIGO did not detect gravity waves and will never detect "quantum nudges." Gravity is the attraction between objects, and attraction doesn't wave. Waveforms collapsing might occur on the chalkboard but Mother Nature isn't having any of it. ALL phenomena are the result of the surface to surface contact between two or more objects- that exist. Always and forever!

As long as quantum looks at "the universe" as comprised of particles, gravity will never be united with light and Strong and Weak nuclear forces. Relativity will never be united with Quantum and scientists will never cross the divide between micro and macro.

Particle scientists will forever be looking for and finding more and more particles. There is even a recent trend of looking for more forces. What? They can't unify what they have now! Yes, and yet they are looking for the "fifth force" which has been hypothesized. This particle and force search is doomed to continue until the cows come home. As long as there are particle physicists, there will be particles and as long as they can not unify the "known" forces, they will be looking for others.

Why do I say this? It is not just because of their scientific method. It is not because of the reification or because they don't define their terms or illustrate their objects so that everyone understands what they are talking about. It's not because of the use of mathemagical equations or because experiments can't work at the "scale of gravitons" or planets. It is not because the size of the accelerator they need would necessarily be so massive that cause it to become a black hole. Well, those are problems, but no, it is because there are no particles and there are only TWO forces. Push and Pull.

It is not a problem of finding a way to unify relativity with quantum. Theorists will never unify gravity with the other forces because one can not possibly explain gravity (pull) with particles.

Before we look at an alternative to their theories, let's do a little history and see how it is that math became the language of science and how this math evolved to the highest level of abstraction even after Niels Bhor warned that math was becoming too abstract. He said: "In the final analysis mathematics is a mental game that we can play or not play as we choose."

Chapter Thirteen - Theoretical Mathematics

To see how we got to quantum madness let's take another tour down particle memory lane. No need to cover all the dates and times and various theories. Anyone can do a Google search and find plenty of sources for this. We've already looked at a few particle "theories" and their equations along the way from Newton, to Chadwick, to Maxwell, to Bhor, to Einstein to Schrodinger to Hawking and Weinberg and Kaku. "Theories" is in quotation marks because a theory is an explanation. Particle physics has no explanations.

See. The book, "The Best of Rational Science: Chapter One – The Rational Scientific Method.

We've already charted the course through four centuries of particle physics, picking up just a few of the highlights along the way. We can't see everything, and besides, the map is not the territory. Reality has little to do with our beliefs about it. Let's dig in just a little bit deeper to see if we can understand how we arrived today at such high levels of abstraction. After all, science took less of a philosophical approach, we are told, when it turned more to the empirical testing of theories through experimentation and Popperian falsification. Actually, what we find is that experiments will generally tend to confirm our beliefs, and beliefs, opinions, and proof have nothing to do with the scientific method.

Doesn't the Rational Scientific Method (RSM) and the Rope Hypothesis (RH) harken back to the days of Democritus, Leucippus, Epicurus and Lucretius? Isn't Rational Science just a type of sophism, a way to convince and recruit through argument? How is the Rope Hypothesis any different than atomism?

Democritus and the atomists denied divine design and intervention in lieu of what some call materialism. A step in the right direction. They saw the interaction between physical bodies, more specifically of the fundamental unit of matter - the atom, as being responsible for all observed phenomena. In that sense, yes,

Rational Science and the resultant Rope Hypothesis is similar to the Greek atomists. They saw the atom, literally "uncuttable," as the fundamental unit of matter. They thought that whatever we see at the macroscopic level is a result of the interactions between atoms.

The atomists applied this "materialism" to ethics, religion and politics which fall under the heading of philosophy. Rational Science uses the Rational Scientific Method, concerning itself only with forming the hypothesis and deriving a theory from that, as it relates to the branches of science under the heading of physics. There is no arm twisting, or show of hands, peer review or Nobel prizes as in today's (mainstream) science. Each individual must decide for themselves and conclude possible or not possible. Rational Science also sees that phenomena are the result of surface to surface contact between at least two objects. At an atomic level, all phenomena are the result of the interaction between fundamental and composite objects.

Indeed, Zeno's paradoxes, as well as relativity and quantum mechanics are laid to waste under the Rational Scientific Method and with the Rope Hypothesis the buck stops at the hydrogen atom. Not only are hydrogen atoms indivisible, but they are eternal, and their motion never stops. RH takes this further, all hydrogen atoms are interconnected by a two strand electromagnetic rope, so there are no discrete atomic particles. Particle physics takes phenomena and reifies it into particles such as photons, quarks and Higgs bosons. Rope Hypothesis provides the architecture and the underlying physical mechanisms by which one can understand light, gravity and the so-called strong and weak forces.

It is because of our limited bandwidth sensory system that we must break everything down into bite size pieces in order to perceive and move about in our world. From the universe of matter and space, to the color red versus violet, we separate the inseparable in order not to be overwhelmed by our senses. Fortunately, our unlimited ability to conceive allows us to understand that there is no physical object space, that matter is interconnected and that the electromagnetic spectrum is a continuum including our limited portion of visible light.

When the materialists of old, still under the authority of the church, turned to math (see Georges Henri Joseph Édouard Lemaître). God creation became Big Bang Creation. Later spirits were replaced with massless point particles, the Authority of the Church was replaced with the authority of Nobel Laureates and Peer Review, and the Ten Commandments of God were replaced with:

The Ten Commandments of Mathemagic

1) Reify verbs into the nouns of reality

2) Avoid defining terms at all costs!

3) Never...nay, NEVER under, any circumstances define reality (exist, object, concept)

4) If forced to define a term, always use synonyms

5) Never answer a direct question with a direct answer

6) Always answer questions with questions

7) Never enter into thing space (thought space is much bigger).

8) When in doubt, invoke the magical words energy, force, waves, or field...charge, black hole, spacetime, parallel U, dark matter, mass, God works in mysterious ways, Quantum Magic, you haven't taken enough Math, it's been proven

9) Always request references, educational background information, diplomas, and peer reviewed literature

10) Never under any circumstances accept any arguments, no matter how rational, from individuals who can not present the previous documentation.

It is very important to understand that mathematics is not the language of science or physics. Math is not even related to physics and has limited use in science. One should not interpret the physical world from abstract mathematical concepts and claim authority, anymore than one can read a Harry Potter novel and claim their interpretation is the authoritative one. While math is a language, it is not the language of science anymore than German, Latin, or English. Mathemagicians can no longer tell the difference between what is a real object of existence and an abstract mathematical construct. The bobble heads believe what they are told without questioning the rationality of it and revere the Rock Stars of science and Physics.

Mathematical equations are irrelevant to understanding reality. Equations can describe phenomena yet explain nothing.

Why rely on a system of mathematical logic whose assumptions (axioms) can only be proven within the system, or requires another system to validate it, as in Godel's Theorem? Popper's falsification dealt a death blow to the scientific method, and placed it squarely in the realm of religion where true and false, right and wrong is dictated by the authorities and duplicated by lab rats to convince and recruit followers.

And what good to explaining is description anyways? We see something happen and we want to explain it. That is what leads us to the RSM where we then remove the observer in order to be objective as possible.

What good did $f=ma$ do for explaining the mechanism of gravity? What good did $c=fw$ do for explaining the phenomena of light? If gravity is warped space or graviton balls, and light is both a photon and a wave, what was learned about gravity and light? BUT, apply a rational, visualizable entity such as an EM rope and these equations only then make sense.

"Mathematics is a place where you can do things which you can't do in the real world." -Marcus du Sautoy

"Young man, in mathematics you don't understand things. You just get used to them." - Von Neuman

Einstein: "You should be able to explain Physics to a barmaid."

Feynman: "If you can't explain it to a 10 year old, you don't know it well enough."

"Mathematics may be defined as the subject in which we never know what we are talking about, nor whether what we are saying is true." - Bertrand Russell

"When we see one body acting on another at a distance, before we assume that the one acts directly on the other we generally inquire whether there is a material connexion between the two bodies, and if we find strings, rods, or framework of any kind, capable of accounting for the observed action of the bodies, we prefer to explain the action by means of the intermediate connexions, rather than admit the notion of direct action at a distance." - James Clerk Maxwell pg. 122 "A Treatise on Electricity and Magnetism" 1873

"Insofar as mathematics is exact, it does not apply to reality; and insofar as mathematics applies to reality, it is not exact." -- Einstein

I say, insofar as illustrated entities explain phenomena, it applies to reality. and insofar as reality can be illustrated it makes sense.

AND finally, we see that the mathematician wishes for us to believe they have the er…uh…answers:

"Science doesn't explain. Science describes." -- Donald Symanek

Another death blow to science was dealt by Popper's falsifiability.

Chapter Fourteen - Karl Popper
The Definition of Crazy

"No rational argument will have a rational effect on a man who does not want to adopt a rational attitude." — Karl Popper

"No hypothesis will have a rational theory if one does not adopt a rational scientific method." – Monk E. Mind

While this is in no way intended to be a thorough examination of the man or his Critical Rationalism (nor did I exhaust the resources available on the subject), a fair analysis can be made if we are to accept Karl's own words as representative of his position. I quite honestly never came to a clear understanding of his Critical Rationalism, or other philosopher's critique of same, and anyways prefer to keep the main thing the main thing, focusing primarily on falsifiability.

Quotes gleamed from the following references:

Stanford Encyclopedia of Philosophy

http://plato.stanford.edu/entries/popper/

Karl Popper and Critical Rationalism

http://en.wikipedia.org/wiki/Karl_Popper

Nicholas Dykes

A Critique of Karl Popper's Critical Rationalism - Reason Papers

Modern mainstream science has incorporated Popper's falsifiability into its scientific method along with induction, deduction, and circular reasoning. It is not difficult to notice that philosophy had, and still has, a great influence on the sciences.

Karl (May I call you that? Thank You!) Karl was critical of induction. Yet, as one of his critics, Nicholas Dykes, fairly points out:

"Collecting disconfirmations and arguing negatively scarcely differs from collecting confirmations and arguing positively. Both are inductive procedures and, as such, have been disallowed in advance by Popper's rejection of induction. The bottom line which CR [Ed. Critical Rationalism] must confront, however, is that one cannot falsify a scientific theory without inference from observed instances. However much Popper may have rejected induction, his own method was in fact dependent upon it."

And

"If our senses are automatically suspect, as Popper maintained, negative or falsifying instances deserve no more credibility than positive or confirming ones."

A critic, Anthony O'Hear, said, "There can, in fact, be no falsification without a background of accepted truth."

Dykes and other philosophers recognize the true/false dichotomy, yet miss the most important dichotomy in science: object/ concept. That's what we can expect when we have priests and philosophers running the science department.

Karl was a mixed bag of incrogruencies. He disagreed with Niels Bhor's instrumentalism and Copenhagen's interpretation of QM (Quantum Magic) yet supported Einstein's realism. Instrumentalism sees a theory as useful in as much as it explains and predicts phenomena; realism says that reality is what it is regardless of what we think it is.

Popper proposed a physical experiment similar to Einstein's thought experiment (EPR) because he saw non-locality as contrary to common sense and wanted to falsify action at a distance. However, his experiment called for counting particles and particles are the bane to both QM and Relativity alike.

Popper stated that "There is no such thing as an unprejudiced observation" and that "our scientific theories must always remain hypotheses." He also said this:

"The quest for certainty is mistaken .... though we may seek for truth ... we can never be quite certain that we have found it"

"No particular theory may ever be regarded as absolutely certain .... No scientific theory is sacrosanct ..."

"Precision and certainty are false ideals. They are impossible to attain and therefore dangerously misleading ..."

If reality is what it is in spite of what we think it is, and "We never know what we are talking about then why not remove the observer from the scientific method of inquiry? Instead, Popper doubled down on testing and experimentation because, he said, we could never know with certainty if our theories are correct, we can only apply probabilities to their correctness.

"Good tests kill flawed theories; we remain alive to guess again." - Karl Popper

MATHEMAGICS and TRUTH

"A statement is true if and only if it corresponds to the facts."

"Our aim as scientists is objective truth; more truth, more interesting truth, more intelligible truth. We cannot reasonably aim at certainty. Once we realize that human knowledge is fallible, we realize also that we can never be completely certain that we have not made a mistake." — Karl Popper

Not only was Popper a philosopher, he was a mathemagician. Karl had a formula for determining the falseness and 'truthiness' of a theory:

"Informative content, which is in inverse proportion to probability, is in direct proportion to testability."

$Vs(a) = Ct_T(a) - Ct_F(a)$,

where $Vs(a)$ represents the verisimilitude of a, $Ct_T(a)$ is a measure of the truth-content of a, and $Ct_F(a)$ is a measure of its falsity-content.

Although Karl's theory of verisimilitude was seen as deficient, and not widely accepted at the time, it was considered central to his "philosophy of science," and today is reflected in the teaching of students that their "conclusions will always be tentative ones."

And a formula for Predictive ability:

[C.P. + E.S.]=U.P.

He saw the advance of scientific knowledge (wait I thought we could never "know' anything?) as an evolutionary process.

He had a formula for that too:

PS1 > TT1 .> EE1 > PS2

It would seem that Popper's demarcation theory (how we separate science from non-science with falsifiability) would invalidate mathematics, physics, and logic as scientific.

Popper's philosophy of mathematics solved the question of how statements of math like 2+2=4 could never be shown false. By his own account, it is not scientific if it can not, in theory, be proven false. How can we learn about reality from it, then?

Simple, says he, the statement "2 apples + 2 apples = 4 apples" has two applications. It is logically true (tautology- axiomatically true) but in reality it may also be falsified.

Karl Popper has the distinction of bringing falsifiability into the scientific method of Popperlar Science.

So, what of this falsifiability? What ever did he mean by that?

"Nothing in the empirical sciences can ever be proven, but it is falsifiable, that is, it can and should be scrutinized by decisive experiments. The term "falsifiable" does not mean something is made false, but rather, if it is false, it can be shown by observation or experiment."

"In so far as a scientific statement speaks about reality, it must be falsifiable; and in so far as it is not falsifiable, it does not speak about reality." - Karl Popper

According to Stanford's Encyclopedia of Philosophy:

"For Popper, a theory is scientific only if it is refutable by a conceivable event. Every genuine test of a scientific theory, then, is logically an attempt to refute or to falsify it, and one genuine counter-instance falsifies the whole theory."

Sounds like my definition for crazy- doing the same thing over and over again expecting a different result.

What about definitions? Glad you asked! Karl Popper didn't like definitions. Apparently he loved word magic!

"One should never get involved in verbal questions or questions of meaning, and never get interested in words. If challenged by the question of whether a word one uses really means this or perhaps that, then one should say: "I don't know and I'm not interested in meanings......one should never quarrel about words, and never get involved in questions of terminology. One should always keep away from discussing concepts."

"Definitions do not play any very important part in science .... Our 'scientific knowledge' ... remains entirely unaffected if we eliminate all definitions."

"Definitions never give any factual knowledge about 'nature' or about the 'nature of things."

"Definitions .... are never really needed, and rarely of any use." - Karl Popper

Poppercock!

"Anyone who refuses to define uses word magic!" – Monk E. Mind

"Not to have one meaning is to have no meaning, and if words have no meaning, our reasoning with one another, and indeed

with ourselves, has been annihilated. - Aristotle
John Herman Randall, Jr., Aristotle (New York: Columbia University Press, 1960), p. 116.

"Science must begin with myths, and with the criticism of myths." - Karl Popper

God and Exist

"I don't know whether God exists or not. ... Some forms of atheism are arrogant and ignorant and should be rejected, but agnosticism—to admit that we don't know and to search—is all right. ... When I look at what I call the gift of life, I feel a gratitude which is in tune with some religious ideas of God. However, the moment I even speak of it, I am embarrassed that I may do something wrong to God in talking about God.

"Natural laws .... do not assert that something exists or is the case; they deny it.

"If we call the world of ... physical objects ... the first world, and the world of subjective experiences ... the second world, we may call the world of statements in themselves the third world." - Karl Popper

Well, if Karl would have only defined the term EXIST, he would perhaps have had a different take on this. Critical to the scientific method at the hypothesis stage is defining one's Key Terms.

Exist: object with location; something somewhere; physically present

I propose to you that science is never about true or false, these are questions to ask your priest or philosopher, not your scientist!

Science is about explaining, and the conclusion is yours to make: Possible or NOT possible.

Chapter 15 - Word Magic

I was in a supposedly "forward thinking" science forum once filled with theoretical mathemagicians. This is part of a conversation I had with a regular member there.

MonkEmind: If someone says that square circles exist and they have the math to prove it, do I need to check their math?

Mathemagician: "If someone had a theory that made useful predictions about the behaviour of reality, and could be used to make cool technology like transistors, and the only way you could get it to work and give those predictions was to assume the existence of hypothetical, mathematical square circles, who are you to call that theory "wrong" or "false"? The universe isn't obligated to be easy for us to understand, any more than it's obligated to be easy to understand for a mouse."

Should I have been surprised to hear this? I wasn't. Listen to the statements of two Rock Starz of Quantum mechanics.

"Everything we call real is made of things that cannot be regarded as real." - Niels Bohr

"Reality is in the observations, not in the electron." – Werner Heisenberg

Science in general, and physics in particular, should be about what is real; objects with location, physically present entities that interact, accounting for all the phenomena whether we observe it or not. Theoretical physicists are magicians, and mystics and spiritualists. If you do not believe this, I will let them speak for themselves.

"Science is not only compatible with spirituality; it is a profound source of spirituality." - Carl Sagan

No wonder a Minister told me that the language of God is math!

"I actually think Deism, the possible existence of a divine intelligence, is not an implausible postulate. And I won't argue against it. It could be. I mean, the universe is an amazing place! So I think the possible existence of a divine intelligence is perfectly plausible and addresses some of the perplexing issues associated with the beginning of the universe. In fact, I should say it more clearly: science is incompatible with the doctrine of every single organized religion. It is not incompatible with Deism." -- Lawrence Krauss

Maybe they should take a cue from a famous musician:

"When you believe in things you don't understand you will suffer... Superstition ain't the way!" - Stevie Wonder

Magical thinking is rampant in schools, colleges and universities. Indeed, it is what is being taught full scale. If persons could remember what they learned as toddlers before school and early on in kindergarten they would perhaps understand the difference between objects and concepts, nouns and verbs, magic and reality.

We begin to recover from the unlearning of school and college by applying the Rational Scientific Method. Rational science defines its Key Terms scientifically; clearly, precisely, unambiguously, non-synonymously, and without circularity or contradiction. Those terms are then used consistently throughout the presentation. Then, there is no question that we understand what we are talking about and no wiggle room after the conference when challenged on our meaning.

"Not to have one meaning is to have no meaning, and if words have no meaning, our reasoning with one another, and indeed with ourselves, has been annihilated. – John Herman Randall, Jr., Aristotle (New York: Columbia University Press, 1960), p. 116.

Now a word from one of our sponsors:

Word Magic V1.1

Tired of being pushed into the corner with scientifically accurate words and phrases? Well, no more, astound, befuddle, and bamboozle your opponents with a working knowledge of the top misleading words in the human lexicon. With words like energy, field, wave, and life you'll never be pinned down again!

With WordMagic V1.1, you can say anything, and it is guaranteed to be true, correct and factual!

You can be the life of the party! You can be unequaled in irrelevancy, fabulously fallacious, and deliciously discrepant. Learn how to impress your family, friends and neighbors with counterfactual vernacular and meaningless jargon.

Whether at the seminar or symposium, conference or confabulation, stun your co-workers, colleagues, and collaborators, with inconcise, indeterminate and unspecific Magical words of misrepresentation and miscommunication.

With WordMagic V1.1 you'll be able to define ambiguously, circularly, synonymously, and inconsistently thousands of authoritative, popular words from our tested and true Word Magic Lexicon.

But Wait! There's more! Order now and we'll include, for no additional charge, our very specially narrated CD version. Listen to the smooth, sonorous voices of folks like, Stinking Hawking and Dick Dawk as they teach you how to dodge and weave through any presentation, debate, conversation or convocation.

And we're still not done! For the next 100 persons who order WordMagic V1.1, we'll include, free of charge, the additional Musical CD, "WordPlay." Relax during an office power nap, or send yourself off to Slumberland listening to these favorite hits: Double Talk, Ambiguity, Dubiety, Incertitude, Tergiversation, Equivication, Polysemy, Double-entendre and Enigma.

WordMagic It's not just for scientists anymore!

We learned from Karl Popper that, "Definitions do not play any very important part in science .... Our 'scientific knowledge' ... remains entirely unaffected if we eliminate all definitions." Theoretical mathematicians claim that "the language of science is math." Yet Einstein said this, ""Since the mathematicians have invaded the theory of relativity, I do not understand it myself anymore."

Quantum mechanic Leonard Susskind, In Defense of Common Sense, The Edge (2005), explains how we got to where we are today, "Where intuition and common sense failed, they had to create new forms of intuition, mainly through the use of abstract mathematics... When common sense fails, uncommon sense must be created."

Since they believe Mother Nature defies common sense, "scientists" have given up any hope of ever understanding how she runs her shop. Father Universe is far too mysterious and unexplainably complex for mere mortals to ever understand Him. Therefore, theoretical mathematicians tell us that the best that science can do is describe.

"Mathematics deals not with reality, but with an abstraction of reality: a "model" of just one aspect of the reality we use it to describe." - Dr. Math

"There is no quantum world. There is only an abstract quantum physical description." -- Niels Bohr

"Mathematics may be defined as the subject in which we never know what we are talking about, nor whether what we are saying is true." - Bertrand Russell

Singularities, massless point particles, blackholes, and extra dimensional realities were a given once space, time, fields, waves and energy were given life and pervaded physics. It is no wonder that these ghostlike entities sprang into existence since that very word itself, the word that should be the foundational Key Term of science and of physics was never defined: exist.

"Existence in the mathematical sense is wholly different from the existence of objects in the real world." -- Kasner & Newman (Mathematicians Extraordinaire), Mathematics and the Imagination

"Everything we call real is made of things that cannot be regarded as real." - Niels Bohr

I've quoted Bhor several times now, and in case you don't know who he is, he is one of the founding fathers of quantum theory. He contributed his principle of complementarity and also proposed one of the atomic models, although supplanted by other models, is still being taught today. Not surprisingly, he said this, "When it comes to atoms, language can be used only as in poetry. The poet, too, is not nearly so concerned with describing facts as with creating images."

The saddest part of this is that the images are reifications, abstractions made concrete, verbs turned into nouns; descriptions of phenomena, not illustrations of physical mediators. As "exist" should be the Key Term of physics, the atom, or some OBJECT with LOCATION in respect to all other objects should be the foundational or fundamental object that comprises all matter. Something physically present should be deemed responsible for electromagnetic, strong, weak, gravitational, or any other forces.

"The next time someone tells you that he believes, or doesn't, in the existence of God, space, time, a number or a leprechaun, just ask him to define the word 'exist.' Then just sit back, grab a beer — and watch him make a fool of himself!" -- Bill Gaede

Emeritus Professor Donald Simanek of Penn State University tells us, "Science doesn't explain, science describes." Nobel Prize recipient Steven Weinberg says, "I don't think mathematics can ever be regarded as an explanation, in itself, of anything, and this is not always understood."

I agree, yet common definitions of science, such as this from Wolfram say, "Science is a systematic enterprise that builds and organizes knowledge in the form of testable explanation and prediction about the universe." Science is also NOT about knowing but understanding. Not about predicting but explaining.

Rational Science Vol. II; Experiments Are They Part of the Scientific Method? Rational Science Vol IV; Chapter Three – Experimenter's Regress, Chapter Four – Knowledge and Prediction

Alexander Unzicker sums it up quite nicely, when he says, "If physicists do not understand the what of their theories, they'll introduce a new particle. If they don't understand the when, then it must have happened right after the Big Bang. If they don't understand the where, then of course it took place in an extra dimension. And if they don't understand the how, they will postulate a new interaction. If they don't understand the how much, a symmetry breaking will soon appear. If they don't understand anything, they will propose strings and branes. And if they lose interest in all understanding, there is always the strong anthropic principle. Things have come to a pretty pass."

We can avoid the word magic by defining our terms and by replacing descriptions with explanations, math with illustration.

"I did a lot of work in the '70s with movies and television, so I understand the difference between telling a story in words and telling a story in pictures. The thing about pictures that you shoot with a camera is that they are very, very literal. So you can't get away with vague descriptions. We'll see in great detail what things and people look like." - Margaret Atwood

Chapter Sixteen - Black Holes From the Beginning of Time
Scientific American July 2017

The next several chapters are critiques of articles from popularization science magazines. These magazines are responsible for bringing science to the average Joe, who is interested but doesn't have time or the "proper education" needed to understand such complicated things as Quantum Physics.

This article discusses the search for hidden populations of primordial black holes formed shortly after the Big Bang in an attempt to "solve the mystery of Dark Matter." The authors hope that the Advanced Laser Interferometer Gravitational-Wave Observatory's (LIGO) gravitational wave detectors will validate their theory of primordial black holes in lieu of WIMPS and MACHOS which have heretofore remained unobserved.

The recent claimed detection of gravitational waves created by the collision of two black holes a billion years ago spurred these sky-gazing theorists on. In their minds, the detection of gravitational waves "proved Einstein's prediction of their existence."

The results did puzzle researchers due to the massive size of the black holes. They were three to four times larger than expected, and the chance of meeting along with the resulting merger has a very low probability given the age of our universe.

Of course, in their mind, this means an even more exotic mechanism of formation must be involved. Hence primordial black holes (PBH) with a broad range of masses are theorized. The nudnicks never stop to consider the irrationality of Big Bang creation, black holes or gravity waves. Cosmologists have piled ad hoc hypothesis on top of ad hoc hypothesis for so long that it never occurs to them they might need to erase the whiteboard and start over with a clean slate. Besides, they have long forgotten (or perhaps never understood) that they live in a fantasy world of abstract mathematical concepts where not a single object can be found.

If you think I am being a bit harsh, then let me quote from the article. "Beyond its detection of gravitational waves, it may be that

LIGO has unveiled something even more extraordinary: black holes that predate the formation of the stars themselves."

Before you jump to conclusions and think the proposal is one of time traveling black holes so that they actually precede the stars that formed them, understand they are just proposing a new type of black hole. Not that this type of rationale is beyond them, just keep in mind that they believe all of matter and space and time began at the Big Bang. So, if black holes are not formed by collapsing stars, what then? Hot, dense plasma.

If PBH theory is "correct" it could solve a number of "mysteries" cosmologists have scratched their heads over for years, in particular, Dark Matter (DM). DM is the glue that holds galaxies together gravitationally. PBHs, massive compact halo objects (MACHOS) and weakly interacting massive particles (WIMPS) are the contenders for explaining the rotational speed of galaxies and why they don't fling stars out like a merry-go-round flings small children out into the park. Come to think of it, the amusement ride may hold the answer to that and also why the outer edge of galaxies rotate at about the same speed as stars closer in; all of the stars are connected by a physical mediator which binds them gravitationally.

Researchers have mapped the clumps of gas in galactic halos and find that they correlate closely with temperature fluctuations in the cosmic microwave background (CMB). Since DM warps space it bends light allowing observers to detect MACHOS, WIMPS or PBHs whichever they are) via something called gravitational lensing. MACHOS and WIMPS have fallen out of favor since none have been directly observed via microlensing, particle accelerator experiments or space telescopes.

The article mentions Hawking radiation and how it attempts to solve an issue of expansion related to black hole size. The author claims that quantum fluctuations could magnify cosmic inflation and "produce particularly dense regions that would collapse to form a population of black holes less than one second after the inflation ends..." These PBHs would explain LIGO's black hole mergers as they would spiral around each other for millions of years "radiating gravitational waves" before finally merging.

The question, "Is Dark Matter Made of Primordial Black Holes?" is like asking if fairies are made from pixie dust!

The rest of the article covers how PBH could answer various questions like the lack of Dwarf Satellite Galaxies, and the origin of super massive black holes (SMBH). The author acknowledges that "other models and explanations are possible..." Well, thank Sagan for small miracles! Now, if only this person would acknowledge that Black Holes are not possible!

Why would I say such a thing? Because whether primordial black holes exist depends on whether black holes exist in the first place, whether or not the Big Bang occurred and whether or not gravitational waves can be detected. They don't, it didn't and they can't!

Black Holes Do They Exist?

Well, that depends on two things: the definition of black hole and the definition of exist. Before defining these terms, let's take a brief look at the varying opinions. I Googled, "Do black holes exist?" and randomly selected these top links. They probably have, might not have, apparently have, cannot have, and have been discovered.

MacDonald Observatory: "Do black holes exist? Probably. Astronomers have discovered quite a few objects that can only be explained as black holes. These objects are dark, so we cannot see them, but they exert a powerful influence on stars, gas, and even space around them. These objects are so dark, dense, and heavy that they must be either black holes or something even more exotic."

NewScientist:
"Black holes might not exist - or at least not as scientists have imagined, cloaked by an impenetrable 'event horizon.' controversial new calculation could abolish the horizon and so solve a troubling paradox in physics."

PBS - Stephen Hawking's Universe:
"A supermassive black hole with 2 billion times the mass of the sun apparently lurks in the nearby Giant galaxy M87. ...stars move

about so quickly that they must be caught in the grips of a massive object. By calculating the size and mass of these objects, the only conclusion seems to be that the center of these galaxies harbor supermassive black holes."

Nature:
Black holes are staples of science fiction and many think astronomers have observed them indirectly. But according to a physicist at the Lawrence Livermore National Laboratory in California, these awesome breaches in spacetime do not and indeed cannot exist."

Harvard:
"They say that truth is stranger than fiction, and it turns out that nature is stranger than science fiction. More than a dozen black holes have already been discovered in our Milky Way galaxy - out of more than a million black holes estimated to exist there."

Why are there uncertain or conflicting answers coming out of McDonald Observatory, New Scientist, PBS, Nature, and Harvard? Why are we left with more questions than we started out with? Physicists have been theorizing and astronomers have been looking for black holes for over a hundred years.

What is a black hole? There are many varying definitions:

Free Dictionary:
"An area of spacetime with a gravitational field so intense that its escape velocity is equal to or exceeds the speed of light."

Dictionary.com:
"A theoretical massive object, formed at the beginning of the universe or by the gravitational collapse of a star exploding as a supernova, whose gravitational field is so intense that no electromagnetic radiation can escape."

Merriam Webster:
"A celestial object that has a gravitational field so strong that light cannot escape it and that is believed to be created especially in the collapse of a very massive star."

So what is it: an area of spacetime, a theoretical massive object, or a celestial object? We are told that a black hole is an area or a region of space with so much mass that nothing, including light, can escape its gravitational pull.

To understand what cosmologists are trying to tell us we need them to define object, spacetime, mass and singularity. There is confusion about the black hole, and it is because of the contradictory, ambiguous terms being used. Area is a concept, and object is that which has shape, mass is a property of matter, and time is a concept relating motion of objects and memory. At the center of the black hole is a singularity. What is that?

That is an important question. The deeper we look, the more questions we have, in a never ending search for what we thought was a simple question looking for a simple answer. Of course, we are told that there are no simple answers. The universe is a strange place and does not conform to our expectations of being logical and rational.

In physics it is crucial to understand what that little word "exist" means, and Einstein rightly points out that whether or not there is an observer has no bearing on what exists. What does it mean to exist? This is the most basic question of science in general and physics in particular. One would rationally think that scientists could tell us. However, astonishingly they cannot! Physics, which is the study of what is physically present, reality, doesn't have a clue! Herein lies the reason for varying opinions on whether or not black holes exist. This is why we get ambiguous or conflicting definitions from mainstream science.

Cosmologists and theoretical physicists live in the make believe world of thought space where abstract concepts and imaginary numbers bump into each other instead of living in thing space where real objects are located.

Don't buy it! Black holes are impossible because singularities are impossible. Zero dimensional point particles are mathematical constructs and a figment of the imagination.

"A black hole is a simple object that has only a 'center' and a 'surface.' " (Universe, William Kaufmann, p. 469)

"[Penrose] showed that a star collapsing under its own gravity is trapped in a region whose surface eventually shrinks to zero size. And since the surface of the region shrinks to zero, so too must its volume. All the matter in the star will be compressed into a region of zero volume...In other words; one has a singularity contained within a region of spacetime known as a black hole." (Stephen Hawking, A Brief History Of Time, p. 49)

A region with zero size? Right. Here are some more black hole absurdities:

Black hole theory says that a black hole has an escape velocity of the speed of light in a vacuum, yet we cannot see a black hole. We are told light cannot escape from it!

Black holes contain an infinitely dense point-mass singularity, yet Special Relativity forbids infinite density as does rationality. So what is a black hole? It is an abstract mathematical concept with no corollary in reality.

What does it mean to exist? To be physically present, that is to exist, is to be an object with a location. Do black holes exist? Obviously not!

Singularities fail conceptually, as zero volume means no length, width, or height- an irrational proposal with no corollary in reality. Infinite density fails even the most basic math as one cannot divide by zero as required by the simple math formula of density, mass, and volume, and infinities cannot exist in reality regardless of what mathematicians do with 'higher math.' Black Hole also fails at the higher level of Newtonian and Einsteinian theory and corresponding maths. Black Hole can only be seen due to so-called gravitational lensing which is circumstantial at best, and then only as an ad hoc presentation in lieu of observation of an event horizon or Black Hole itself.

Einstein's Special Theory of Relativity clearly prohibits infinite densities. Karl Schwarzschild attests to this fact, proving it in his paper on point mass ('On the Gravitational Field of a Mass Point

According to Einstein's Theory'). This was also confirmed by Leonard Abrams' paper 'Black Holes: The Legacy of Error.' Not only this, Einstein denied the possibility of black holes multiple times before his death in 1955. Both Einstein Relativity and Schwarzschild's solution theories forbid infinite densities.

So does common sense and rationality. Although math can postulate infinite densities in abstraction, reality is having none of it. While Hilbert can build hotels in thought space with an infinite number of rooms, the world of reality does not comply. Though Zeno in his Dichotomy Paradox can halve a distance infinitely in abstract mathematical equation space, one cannot walk half way to a brick wall indefinitely and certainly cannot halve their distance an infinite number of times. All who have tried end up smacking their foreheads on the wall.

Furthermore, later work in the 40's conveniently ignored relativity and erroneously posited an "infinity of spacetimes differing as to the limiting acceleration of a radially approaching test particle." In other words, Hilbert substituted a variable with a scalar invariant transforming the coordinate location of a point mass. Because of the error, the point r=0 becomes a two-sphere invalidating Hilbert's assumption.

There are no known solutions for Einstein's field equations for two or more bodies, and yet proponents of Black Hole allege multiple masses interacting with each other and with matter. The principle of superposition applies to Newtonian masses but not to General Relativity, so Newton's escape velocity cannot be used in an expression relating to a universe containing only one mass. Einstein's theory, as Schwarzschild shows, pertains to one mass. In other words, Newton's theory contains two masses and superposition. However, r=0 contains no bodies, and so therefore cannot accommodate superposition. Also, Newton's escape velocity has no relevance in a universe of only one mass as required by Einstein and Schwarzschild's theories.

Gravity

Let's take a look at General Relativity's (GR) proposal or gravity. General Relativity is founded on geometry. Geometry is supposed to be about objects. Relativity tells us that the three dimensions of reality combine with time and form the fourth dimension known as spacetime. Yet, neither time nor space are objects. The three dimensions of length, width and height are the only mutually orthogonal directions an object can face. These three are perpendicular to each other, an imaginary time line can not be perpendicular to anything. Not to mention there is no fourth perpendicular direction an object can face.

GR describes gravity as warped or curved space. A heavy object like the sun pushes down on space resulting in something called a "gravity well," on which the earth rolls down, and around, and around, and around the sun it goes.

"Matter tells space how to curve. Space tells matter how to move."
- *Wheeler*

Think how irrational this is. Matter curves space, and space moves matter. What is gravity, again? It is heavy objects weighing down space causing space to curve forming gravity wells that cause matter to move. Circular! If the sun weighs down space because of gravity, then gravity has nothing to do with this "Fabric of the Cosmos."

Gravity waves

We cover gravity waves in the article Gravitational Waves From novelty to science

For now, let's just say that it is impossible to detect gravity waves because a wave is what something does, and gravity is attraction between two bodies, ergo gravity waves are the waving of the attraction between two bodies. How does attraction wave? It's not only grammatically incorrect, it is irrational.

Dark Matter

If you have ever played Tether Ball, then you have noticed that as the rope wraps around the pole and gets shorter, the ball rotates faster and faster. The Caballero with longer ropes on his bolo has to spin it around more times to get up to the same speed as a Caballero with a bolo that has shorter ropes. All of this is because speed is a function of the distance traveled around the axis of rotation (center of gravity). The earth takes less than half the time that Mars takes to orbit the sun because it has less than half the distance to travel.

However, this is not the case for the outer stars of a Galaxy. The stars on the edge of the Milky Way, for instance, appear to be traveling at roughly the same speed as the stars closer to the center.

Since this falsifies Newtonian-Keplerian prediction and General Relativity, the theorist invented a type of invisible matter (an enormous amount of mass) to account for the phenomena. We are told that since this matter does not emit or reflect electromagnetic radiation in large enough quantities to be seen, it must be inferred from the gravitational effects on matter that can be seen. No one ever questions why the Dark Matter is transparent (that is, one can observe stars right through them).

Big Bang

The Big Bang Theory says that all matter, time and space were created at a single moment in er...time. Big Bang is a creationist proposal of massive stupidity. There is no rational way to explain how something can come from nothing, or how matter and space had a beginning. To call for a beginning to matter, space and time is religious creationist nonsense. Come by The Rational Scientific Method facebook group and let's discuss this with the luxury of detail!

Chapter Seventeen - Gravitational Waves
From novelty to science
Astronomy November 2017

Yet another article takes a look at LIGO's claimed detection of colliding black holes. It claims this "opens the window to an exotic new realm of astrophysics." I'd have to agree with that assessment, as nothing is more exotic than this: "The wave's peak power briefly exceeded the total energy output of the entire observable universe 50 times over..." So, naturally, the very sensitive equipment was able to detect the "infinitesimally tiny portion of its energy" that passed through the earth as the "wave radiated" 1.3 billion light years in all directions.

Although the article covers what they consider as the significance of LIGOs experiments in regards to understanding large black holes, black hole binaries and spinning neutron stars, let's take a look at LIGO itself. The engineering marvel, whose 2.5 mile long arms were "stretched and compressed" in the two LIGO facilities, allows detection of gravity waves with a "precision equivalent to measuring the 4.2 light year distance to Proxima Centauri to the width of a human hair."

Lasers are fired down the two arms and then combined to form an interference wave pattern. Scientists analyze these wave patterns and calculate tiny variations in distance the laser light travels reportedly caused by "a passing gravitational wave." LIGO enthusiasts hope that these observations will help to determine which type of black hole predominates; those that formed separately and then came together, or those that were formerly binary stars, and became black holes together.

LIGO wants to test one of relativity's predictions that all frequencies travel at the speed of light. Of course, they never stop to consider that there are no black holes and no such thing as gravity waves. Let's humor them for a moment and assume there are. Rational scientist Wladimir Reiswich has this to say: "I'm interested in this stuff from an engineer's perspective. My field of expertise is microwave engineering and many techniques employed in the visible spectrum are also applicable in the lower frequency bands.

"Just to put in perspective what the people at LIGO have claimed to have done let us scale everything up a bit. If we stretch the 4 km long resonator to the moon (for the simplicity of the argument say 400k km) we'll end up with an alleged shift of the mirror of just the diameter of a hydrogen atom.

"The method employed by the LIGO interferometer is that of destructive interference. The beams have to have a phase shift of pi or 180° at the detector in order to extinguish each other. There is no way of measuring the phase shift directly so they have to rely on the intensity measured at the detector in order to infer it. In a perfect world no light would reach the detector and only by moving the mirrors the intensity would vary.

"However, as one might expect there are several problems that introduce noise into the signal. The first is that a beam splitter is not perfect, meaning it won't redirect 50% of the power of the laser beam in both directions. If the beams arriving from 2 different directions at the detector are not exactly of the same amplitude they don't extinguish each other and introduce some noise.

"The nature of light is that if it hits a polished plate (such as a silver mirror) it pushes on it. This principle was used in order to calibrate the mirrors. With the 100k watt laser employed by LIGO this would mean that just a variance of 10^{-12} of the intensity the mirror could be pushed far enough to generate the same signal.

"Thermal noise is always a concern and there is absolutely no way to eliminate that unless you cool the entire apparatus down to liquid nitrogen levels. LIGO claims that the signal to noise ratio (SNR) was at 20dB which is a factor of 100, which is huge. With the signal that small the brightening of the beam would be in the range of 10^{-18} Watt implying that the noise level would be in the 10^{-20} Watt range.

"Even if I concede that the detector and ALL SUBSEQUENT amplifiers and network don't generate any noise and that the detector/filter has an exceptionally low bandwidth (which is unrealistic to say the least) this is still off by a factor of 10."

How do they analyze these patterns? According to Bailes, Thrane & Lasky:

"The gravitational waves are detected by fitting one of a large number of theoretical "templates" to the data. These templates model how the detectors will react to the passing waves from different mass black holes."

Stephen J. Crothers, cosmologist, clarifies the process:

"The actual number of templates in the LIGO database is 250,000. All of these are constructed by incorrectly assuming multiple black holes, and then applying to them different parameters such as mass, speed, rotation, etc. None of these templates have any physical meaning because they are built upon false hypotheses.

"LIGO first detects a generic "signal." Its massive computer power then goes to work to find a line of best fit to this alleged signal by comparison with its database of mathematically produced patterns. The line of best fit so obtained is then purported to identify a physical occurrence – in this case as with the past two, two black holes merging to form one black hole, with emission of gravitational waves.

"With such a large set of arbitrary templates and such computing power, it is possible to find a line of best fit for any pattern of disturbance, even for Whistler's Mother. There is no upper limit on the number of template patterns that can be arbitrary manufactured and included in the template databases. No doubt the LIGO Team will augment their database with evermore patterns, as is their wont in general."

The article on gravitational Waves informs us of the corroborative effort between LIGO and Europe's VIRGO interferometer. With just the two interferometers, the location of a source of gravitational waves can be pin pointed to a precision of 100 square degrees. Adding the third interferometer allows triangulation of a signal to 10 square degrees. Hopefully, this year, Japan's two new interferometers, KAGRA, will join with LIGO and VIRGO offering an even greater precision.

LIGO hopes to detect other objects besides black holes and neutron stars; cosmic strings and supernovas. Supernovas would release neutrinos on the order of 10^{58}. Earth's multiple neutrino detectors would detect them at the same time as LIGO detects the supernova giving scientists "a window on exactly what was happening in real time."

Whatever they are detecting now or hope to detect in the future, it is not gravitational waves from colliding black holes, and black holes are not created by a Big Bang.

Here are a few tidbits from Mr. Crother's which are covered with great detail in his many papers available on vixra:

A black hole is eternal and a Big Bang U is not, therefore a BH can not be part of a BB U theory. A black hole is infinite and a BB U is not therefore a BH can not fit in a finite BB U. There can be no parallels between BH laws and Newton laws because there can be any number of masses in Newtonian physics but only one BH in Einsteinian physics.

Here Bill Gaede adds his comments to one of Crother's papers on black holes.

"General Relativity has to do with the macro world of planets, stars and galaxies. However, all equations invoke the gravitational constant G which is determined via the Cavendish Torsion Method: a micro world experiment performed in labs here on Earth. Therefore, General Relativity equations have no business using the gravitational constant G unless the math can be extended to describe micro scenarios as well.

Here's a little taste of how Mr. Crother's destroys theoretical mathematicians using their own language of math, but dumbed down for us mere mortals:

"At the black hole singularity, volume is 'zero,' density is 'infinite', and spacetime curvature is 'infinite' (hence gravity is infinite).

"Now one need not be proficient in math to see that according to the rules of math and determining density, this is impossible. For to divide by zero violates even the most basic math.

"density=mass divided by volume.

"Dividing by zero is called an indeterminate form. One can not determine a value for density if one must divide mass by zero.

"In physics, Density = Mass/Volume. In symbols, $D = M/V$. Now $M > 0$ for any physical object. Since $M > 0$, $V > 0$, because no mass can have zero volume (at the very least it has an atom). Hence in physics, D is always finite. From a purely mathematical perspective however, if $M = 0$ and $V = 0$, then $D = 0/0$. The mathematicians call this 'indeterminate.'

"According to the pure mathematicians, if $M =/= 0$, then the calculus takes limits: then the limit of D as $V \rightarrow 0$ (i.e. as V approaches zero) does not exist! If $M =/= 0$ and $V = 0$ then $D = M/0$. According to the pure mathematicians, M/0 is 'undefined'. One cannot divide any real number by zero - it is 'undefined'. But cosmologists permit division by zero, and call it 'infinity'. But infinity is not a number, and so it cannot be used to do arithmetic or algebra etc.

"Giving 'infinity' a symbol does not make it a number. There are symbols for male and female, but they are not people. Use the symbol o-o for infinite here. If it is a number then we can do arithmetic and algebra with it. According to cosmologists, they can increase or decrease o-o by a finite amount and still get o-o. Now 1 is a finite amount. Then o-o + 1 = o-o.

"By the rules of algebra, subtracting a number from one side of an equation requires subtraction of that same number from the other side of the equation, to maintain equality. Hence o-o + 1 - o-o = o-o - o-o. A number subtracted from itself always produces zero. Hence o-o - o-o = 0. From the foregoing, we get 1 = 0, which is false. Replace 1 by any number n =/= 0. Then we get n = 0, which is false. Thus, the assumption that infinity is a number is false.

"Cosmologists divide by zero to get infinity. This is how they get their black hole event horizons and black hole singularities, and their big bang singularity. Now consider Special Relativity. In SR, density of a mass is inversely proportional to the square of the speed v of the mass. Then the limit of D as v --> c (the speed of light) does not exist! Thus, by SR, infinite density is forbidden. Now General Relativity cannot violate Special Relativity. Thus General Relativity also forbids infinite density. Nonetheless, cosmologists claim that their finite black hole mass is concentrated at its singularity where volume is zero, density is infinite, and gravity is infinite. Such a notion violates physics and mathematics. They do the same for their big bang singularity, which they claim was of zero volume, contained all the mass of the universe, and was infinitely hot.

"Also, what is temperature? According to the physicists and chemists, temperature is the manifestation of the kinetic energy of particles, be they within solids, liquids or gases. So just how fast must atoms or fundamental particles be moving to present 'infinite' hotness? And since the big bang singularity volume was zero, where was the space to contain the particles and also for them to race about in order to produce a temperature in the first place? Cosmologists don't even understand the meaning of temperature either. Cosmology is not science - it is nothing but mysticism, irrational imagination, and superstition, masquerading as science, perpetrated by money-driven nitwits who have no scientific acumen."

For more from Stephen J. Crothers: Crothers, S. J., Black Hole and Big Bang: A Simplified Refutation,

http://vixra.org/pdf/1306.0024v2.pdf

Crothers, S.J., A Critical Analysis of LIGO's Recent Detection of Gravitational Waves Caused by Merging Black Holes, Hadronic Journal, n.3, Vol. 39, 2016, pp.271-302

http://vixra.org/pdf/1603.0127v4.pdf

What is gravity?

Gravity: the attractive "force" (pull) between all objects; waves: what something is doing; I wave my hand, the ocean waves, and the amber waves of grain.

Gravity waves: the waving of pull

This is not just a matter of semantics (although language is important, and precise definitions of words are necessary for communication). We are being told that LIGO detected the waving of pull between two colliding non-existent, impossible, black holes. Since there are no black holes, there are obviously no gravitational waves caused by black holes colliding.

There is no such "thing" as a gravitational wave. What is waving? Gravity? According to NASA, "Gravity is the force by which a planet or other body draws objects toward its center. The force of gravity keeps all of the planets in orbit around the sun." What is force? According to wiki, "In physics, a force is any interaction that, when unopposed, will change the motion of an object." So how does an interaction wave? It can't. Only objects can wave, or DO anything.

Is there a rational explanation for gravity? You bet there is. Bill Gaede's Rope Hypothesis.

Rope Hypothesis provides the physical mechanism by which Thread Theory explains Gravity and Immediate Action at a

Distance. All atoms are connected by a two strand rope. The tension between objects is the net result of all ropes. This is the force we call Pull (gravity). http://vixra.org/pdf/1205.0015v1.pdf

Chapter Eighteen - Mass and Energy

What ARE They?

Before we critique a few more articles. Let's cover two of the magician's favorite magical words.

We discuss this in the series of articles entitled. "Mass" (In the Files section at Rational Scientific Method fb group and also in my book Rational Science Vol. I).

Mass is said to be equivalent to energy by relativists. Let's see if we can figure out what they mean.

Mass is the quantity of inertia possessed by an object or the proportion between force and acceleration referred to in Newton's Second Law of Motion. - Physics.About

Huh? Either possessed by an object or a proportion between concepts? Well, that was what someone had to say about what he thinks Newton had to say about mass. Maybe we should ask the father of Relativity.

"It is not good to introduce the concept of the mass $M = m/(1-v2/c2)1/2$ of a body for which no clear definition can be given. It is better to introduce no other mass than 'the rest mass' m." - Einstein in a letter to L. Barnett in 1948

Einstein tells us it is one thing to use this first definition of mass (as being inertia) in F=MA, but not in E=MC^2.

Whatever does he mean? Clearly he understood it as a concept. Maybe if we define all the Key Terms in the definitions we can reach some kind of understanding of these enigmatic words: mass and energy.

What is inertia?

"Inertia is the name for the tendency of an object in motion to remain in motion, or an object at rest to remain at rest, unless acted upon by a force." -physics.about

"Force is a quantitative description of the interaction between two physical bodies, such as an object and its environment.

"Force is proportional to acceleration. -physics.about

And

"Acceleration is the rate of change of velocity as a function of time. It is vector." -physics.about

And

"Velocity is a vector measurement of the rate and direction of motion or, in other terms, the rate and direction of the change in the position of an object." -physics.about

So using those exact definitions in one sentence we get this:

Mass is an amount of a proportion of an interaction and a rate of change of the change of the rate and direction of the change in position of an object as a function of time. In other words, mass is the movement of motion of an object over time. Clearly the mathematician confuses concepts and objects and moves motion around with his pencil in his equations.

And what about this mysterious word "object" we have been using? We used About.physics for the other definitions, but it has nothing to say about this word object, so lets use Oxford's Dictionary: *a material thing that can be seen and touched.*

Finally, we get to motion: "In physics, motion means a continuous change in the position of a body relative to a reference point, as measured by a particular observer in a particular frame of reference. In other words, any physical movement or change in position or place." - Wiki Answers

"Newton's second law of motion can be formally stated as follows:

"The acceleration of an object as produced by a net force is directly proportional to the magnitude of the net force, in the same direction as the net force, and inversely proportional to the mass of the object." - PhysicsClassroom.com

So, here we are back at the beginning, and we are no closer to understanding any of it. At least relativists are being consistent in one respect; mass, object and motion require an observer.

We see, in the articles on mass in the Files Section at RSM, that the equations for mass are circular. We see very clearly here that the definitions for mass are also circular. We also note that, by definition, an observer is needed for mass, object and motion.

But if every"thing" has mass and mass is equivalent to energy, then don't we need to understand what is meant by energy?

As a child my science teacher told me this:

Energy is the ability to do work. Energy is mathematically quantified as follows: 1 J of "energy"= 1 W s = 1 kg m^2/s^2

Energy relates matter, its motion, the distance it has traveled, and how much time it took for it to get there. Energy is not a thing. It is relational, in other words, a concept.

Maybe the definition has changed. Let's see: "In physics, energy is a property of objects which can be transferred to other objects or converted into different forms. The "ability of a system to perform work" is a common description, but it is misleading because energy is not necessarily available to do work." - Wiki

OK, I can transfer my energy to a rock by picking it up and raising it above the ground, or throwing it through a relativist's window.

But what does it mean when we say a rock sitting on the ground has potential or stored energy? How can a rock have, or store energy if it is a concept? I store potatoes on a shelf under the kitchen counter. AND how do I transfer energy to the rock by picking it up or throwing it?

From Wiki: "In physics, potential energy is energy possessed by a body by virtue of its position relative to others, stresses within itself, electric charge, and other factors."

What is kinetic energy?

Kinetic energy is energy that a body possesses by virtue of being in motion.

So, I transferred my energy by being in motion (throwing the rock). Since the rock was in motion its potential energy, while just sitting there on the ground, became kinetic energy because it was now moving. But what was actually transferred? Stresses within me, electrical charge, other factors? When the rock went through the relativist's window, the rock transferred its energy to the window which broke into many pieces and fell to the floor. Unless Mrs. Relativist adds her energy to the broken glass's potential energy, the glass will lie there forever because Mr. Relativist never does physical work.

The sun provides energy to a spinach plant which needs it to grow. Solar energy is transferred to the plant and by virtue of photosynthesis the sun's energy becomes sugar in the spinach leaf, and the sugar becomes ATP through cellular respiration. A rabbit comes along and eats the spinach leaf. Now the solar energy which became sugar can become rabbit energy. The rabbit hops around looking for more spinach to eat in Mrs. Relativist's garden.

Sun, spinach, sugar, ATP and rabbit are the objects. But energy is the transfer of motion. Motion is two or more locations of an object. Transfer is an act of moving something to another location. Huh? So energy is moving motion. Grammatically incorrect and irrational!

At least the phiz whiz is consistent. AND we see that mass which is motion in motion, and energy which is moving motion are the same thing!

Chapter Nineteen – TIME
Popular Science
September/October 2017

The entire issue is about the "Mysteries of Time and Space." Actually, it's really all about "time." Even some of the ads are about time (watches). The articles cover a wide range of temporal topics, from caribou punctuality to the best time to eat, sleep and be creative. Although interesting and engaging, we never discover what time actually is. The enigmatic word "time" is never defined.

From the first page, on "how long it will take you to read this issue from cover to cover," to the last page, "I Wish Someone Would Invent...A pill that stops aging," the word time and its many permutations are used dozens, maybe hundreds of times. We'd give you the actual number, but we haven't the time to count them all.

The word "time" is one which we all use in at least one of its forms on a daily basis. There are 52 synonyms and 163 other time related words at thesaurus dot com. Here's their definition:

Time: *noun*, temporal length of event or entity's existence, period

Rational scientists reject this definition as circular. A scientific definition never uses the word or synonym of the word in its definition. Temporal is "relating to time." Period is "a length or portion of time." What have we learned?

Time is not a word that can be used as a Key Term in science. We "kill the observer" in order to be objective. Time places the observer front and center in its presentation. Engineers measure, scientists explain. We will cover the different angles or perspectives of time covered throughout the magazine and then, give a scientific definition of time.

The first article, entitled Head Master - Your Brain: Time Machine, discusses how the brain is like a clock. Neuroscientists don't "fully understand" how the brain does this, but they do know that different parts of the brain are involved in different timing tasks. All the different areas of the brain work together "to shape our temporal perception."

The Supplementary Motor Area is like a stopwatch which starts counting (neurons fire in succession) when a particular stimulus from the environment, such as the smell of French fries, triggers it. This "information" is passed along to other areas of the brain.

Estimating time, intervals, duration, etc. would be impossible without short term memory which is housed in the right inferior frontal cortex. The sequential order of events may be sorted in the hippocampus; at least that area of the brain is active during this process. A dancer or a drummer's cerebellum is probably responsible for keeping the rhythm or the beat by coordinating muscle movement, and the Basal ganglia uses a process called delayed discounting to tell you how long to wait for a payoff in the form of that piece of chocolate cake.

"Teach AI When to Say Hi" underscores the importance of timing for a robot bar tender. Oregon State is "teaching" AI to "understand how humans perceive time." "Not in Sync – Consider the Caribou" covers the importance of their internal clocks to keep track of seasonal changes. In "Different Strokes – Which Way to Tomorrow," a linguist reports on the different ways that various cultures perceive time. He rightly points out that time is abstract, and different cultures use their own metaphors "to imagine the fourth dimension." "Mind Your Time – Where Does the Day Go" discusses how little events such as cleaning the toilet or standing in line at the grocery store are less memorable but make up most of our typical day.

"Tock Ticker – A Brief History of Time (Keeping)" covers the advances from water clocks and sun dials to atomic clocks. "It's All Relative" compares the difference in speed of Hussain Bolt, a tortoise and a hare, etc., showing that "the concept of a minute is kind of meaningless."

"A Twinkle in Time" gives the perspective of time in terms of the distance light travels from various celestial objects, in effect, allowing us to see what they looked like eons ago. "The Man Who Would Kill Your Holidays" is a professor of economy that wants your birthday to fall on the same day every year. "Wait a Second (and a Half)" tells us how the Swiss Federal Railway keeps there trains on time. "Digging up the Past" is about the "lure of treasure hunting." "Slow down the World" is a review of cameras and their ability to fit ever more frames into a second effectively slowing down or speeding up time.

"When did it All Begin" is about geological evidence for the beginning of life on earth. "Watching the Clocks – Five experts obsess over lines, places, faces and noses to understand how every second shapes our world and how our minds shape every second."

"Elisa Felicitas Arias" is about the Director of the Time Department, a division of the International Bureau of Weights and Measurements. Although she is responsible for keeping all our clocks in sync, she says that there is no "perfect time." "People say the UTC is the international reference for time, but in fact, UTC is just a piece of paper."

Matias Duarte is Google's VP of Material Design and is responsible for making our on-line experience of wait times seem to pass quicker. "The real constraint is human perception." Alexandra Horowitz studies how dogs perceive time using their sense of smell. Psychologist Sylvie Droit-Volet says "Time is plural. We have several clocks attuned to the rhythms of our daily life." Richard Larson looks at what is fair about waiting in line.

"Life Times" discusses how long a year is, comparatively, to a tortoise, a whale, a glass sponge and a seagrass meadow. There are other articles about how your schedule is killing you, about a watchmaker, and how a stroke wiped out a woman's sense of past and future but the article we will discuss with the luxury of detail is entitled "But Wait...What Came before the Big Bang? – Not everyone thinks the universe had a beginning."

The authors suggest that scientists used to think that the universe was timeless, that there was no beginning, and this was easier than trying to "figure out what this would mean, let alone when it would be." The article covers several alternative theories to this cosmic "question" and it is sprinkled with illustrations in the form of 60's psychedelic black light posters because, they say, the "concepts are trippy."

One alternative, they call the Flowers of the Multiverse, says that since there is so much evidence pointing to a beginning, cosmologists envision many individual universes. Each universe has its own beginning and end, but the multiverses are ageless. "Infinite possibilities yield infinite results." Each universe is governed by its own rules. Somehow they see this as an explanation for life "itself." One Big Bang doesn't seem enough, without luck, to produce the conditions for life.

One variation on the Big Bang Theory has everything "mashed into a single ball of energy smaller than an atom" before it "burst outward." The question then is, "What came before?" Stephen Hawking proposed that perhaps there was no clear starting point or big bang (I guess it is fuzzy). He said that asking what came before is like asking what is north of the North Pole. "Time, as we define it, looses its meaning as the universe shrinks down. " Hmm? I wonder how "we" define it.

Another alternative is the Steady State Theory. If there was no Big Bang explosion blowing everything outward in all directions, then how do we explain the observation of galaxies "shooting off into the distance?" Hoyle and others proposed that new matter is being created everywhere and everywhen, pushing the "older stuff" out of the way. The discovery of the left over radiation (cosmic microwave background) "cooked" this idea.

Paul Steinhardt, et all, wonder why there has to be a single Bing Bang at all. What if there are many Big Bangs happening all the time? What if our universe is in a fifth dimension where "all of space and time" lives on a four dimensional surface called a brane which sometimes collides with another universe's brane." Yeah, man, what if?

I remember black light posters and the trippy psychedelic days of the 60's so I understand where these cosmologists get these ideas. But, is there any validity to them? No, there is not. At least nothing we can parse from their language. There are many "alternatives" because there are many different definitions of their Key Terms: Time, space and exist.

We'll cover the terms space and exist in the articles on black holes and gravitational waves, so for now let's look at, "time."

What is time, anyways? Popular Science writes about how humans and even dogs perceive the "passage of time" differently. Neurologists seem to understand that time is something that the brain does. That this is the brains way of ordering things and depends on external cues and memory. The keeper of time at the Bureau of Weights and Measures told us that the Universal Time Code is "just a piece of paper." One article said that the concept of a minute is meaningless. A linguist recognizes that cultures perceive time differently, and a psychologist says that time is plural. Clearly, time has different meanings and different significance to different people and cultures, but scientists may have the strangest ideas of all.

Cosmologists believe that time had a beginning along with space and matter, as though time is something physical like a star or a planet. Although Hawking is quoted as saying, "…time as we define it…" there is no single definition for time that can be used consistently across different scientific disciplines or throughout various alternative theories. In fact, in his book, A Brief History of Time, there was not a single definition of that Key Term.

"A continuous, measurable quantity in which events occur in a sequence proceeding from the past through the present to the future" – dictionary dot com

A quantity of what?

"*Time* in physics is *defined* by its measurement: *time* is what a clock reads. In classical, non-relativistic physics it is a scalar quantity and, like length, mass, and charge, is usually described as a fundamental quantity." – Wiki

A measurement of what?

"In physical *science, time* is *defined* as a measurement, or as what the clock face reads." — www.reference.com

A reading of what?

"It is not surprising that the *meaning* of the word *time* cannot be distilled into a neat verbal *definition*." – Britannica

Why not?

We sort of expect dictionaries to vary in definitions because after all, these are just the popular usage of the words, and these vary over time and locale. We understand that popularization science magazines will also vary because they are written as entertainment for the general public and don't need to be so precise. But, asks a scientist, like Sean Carroll, what time is and, well, here is some of what you'll get.

Wired.com: Can you explain your theory of time in layman's terms?

Sean Carroll: "I'm trying to understand how time works. And that's a huge question that has lots of different aspects to it. A lot of them go back to Einstein and spacetime and how we measure time using clocks. But the particular aspect of time that I'm interested in is the arrow of time: the fact that the past is different from the future. We remember the past but we don't remember the future. There are irreversible processes. There are things that happen, like you turn an egg into an omelet, but you can't turn an omelet into an egg. And we sort of understand that halfway."

In science, a definition is a limitation or restriction on the use of a word that the person using it must define in order for his or her audience to understand what he or she means in the presentation. A scientific definition does not call upon an observer, it is not synonymous, it is not circular and it is not contradictory.

Let's see if we can define "time" scientifically.

Time: two or more locations of an object plus memory: motion plus memory.

Does this not fit well for everything that has been discussed with the exception of the Big Bang stuff? If, as one of the articles points out, the brain is like a clock. What is a clock? A clock is a device which compares the motion of one object against the motion of another object. In this case, the motion of the earth around its axis and its orbit around the sun.

The following is an excerpt from my book, A Curious Chain of Events: There is no such thing as time in physics. Time is a scalar quantity used for measuring movement. It doesn't slow down, speed up, dilate, contract, bend, warp, shrink, or expand, any more than a meter shrinks or expands. Only the object being measured can speed up, slow down, shrink, or expand.

The word time should not be part of the vocabulary of science, or of physics. We cannot explain any phenomenon of nature with the word time. Time is descriptive. Time doesn't exist. Time is a human conception.

After all, your entire existence revolves around the concept of time. Just try uttering a single sentence without referring to time: today I'm going to ..., call me later, when I was ..., how long will it ..., and so on. Time is so innate, ingrained, and intuitive to your thought processes that you don't even give it a second thought. The human capacity for perceiving time has to do with your sophisticated memory, which is a prerequisite for perceiving time. After all, a huge chunk of what you perceive as time, the past, is only made possible by memory. Without a good memory the past is inconceivable.

A rich collection of memories, however, could actually be a mentally debilitating liability without some means to categorize those memories and make some sense of them. Chronological ordering is certainly a useful way to categorize memories, thus necessitating the concept of time. Time is merely an aid to thoughts that require memory.

You are living objects in motion, not in time! The concepts within your human mind's reality have no physical collateral, and hence do not coincide to anything within the actuality of the universe.

Time = object motion + memory

Time is a dynamic concept, a scalar quantity necessarily requiring the measurement of motion of an object. Without sentient observers with memory, there is no time!

Time is not a noun of reality. Time is not real. Time does not exist. Time is an abstract concept, specifically, a verb conceived of by man.

Hawking wrote a book called A Brief History of Time, and nowhere will you find the definition of the center of his dissertation. He states that while he has no evidence of time travel, it is theoretically possible, and that it's possible that man has been visited by more technologically advanced people from the future!

If time has to do with physics, then hopefully it would be defined consistently and unambiguously alongside "object" and "exist," don't you agree?

Time is a concept after all - no different from the concept of a meter or a kilogram, which necessarily requires at least 2 objects to be conceived of and defined. We are living objects in motion, not in time! The concepts within your human mind's reality have no physical collateral, and hence do not coincide to anything within the reality of the Universe.

Time is not some ethereal property or entity, but rather it is just a defined metric like length and mass. In fact, time is simply the scalar metric of motion. And the limit of motion is also the limit of time. So you build an instrument to do the simple task of demonstrating a standard uniform motion, and you cause this instrument to record its accruing motion in some manner which you calibrate to mimic the Earth's rotation. Since this instrument accomplishes no other task but to move uniformly, you call it a clock and say it is measuring time even though it is not!

Our motion-clock is not measuring time; rather it is measuring motion - its own motion. That is why it confuses you. You are taught to think that it is measuring something separate from its own internal motion. After all, you use standard lengths to measure and compare all other lengths, and you use standard masses to measure and compare all other masses. Likewise, you use standard motions, such as the Earth and Sun, to measure and compare all other motions. You do not use clocks to measure indefinable concepts. As uniform cyclic motions occur and re-occur, you identify their rhythmic nature and regulate your motions, i.e., your lives to them, insufficiently aware that here on Earth it is only the Earth's rotation and revolution motions which define your days and years.

There are no absolute metrics in the universe. Every measured quantity is relative to a finite standard that is pre-defined by man, but not measurable. By definition, a Standard Metric is simply one unit of Mass or Length of

Motion. You determine how finite something is by comparing it with your defined unit of finite length.

Similarly, for motions you compare motions and changes in motion in a relative sense of change and not in any absolute sense. There are no absolute scales of measure. In calculations involving time as motion you use the comparable motion of the Earth's rotation camouflaged in clocks, while in your descriptive verbiage about time you use some confusing ideas that suggest a universal property or dimension. This is utter nonsense called the Fallacy of Reification.

Time is a concept that is defined by change, or cause and effect. Time is a verb. No physical object is subject or physically dependent upon the concept of time. Time is a figment of the imagination of a living entity and nothing more. Time is not a part of reality. It belongs exclusively to the religion of mathematics.

Time is a word that serves the interests of mathematics because math is the study of dynamic concepts. It has no bearing on issues concerning physics, the study of existence. In math, they stretch and warp and bend seconds because they don't understand the difference between an object and a concept. In the reality of science, we stretch bubble gum, warp hammocks, and bend molten iron rods.

If there were only the Sun and the Earth in the entire universe and they had no motion, then humans would not be able to conceive of time as you use it today.

One side of the Earth would always be day and the other always night. So humans would need to devise a different "counting mechanism" in order to segment before and after events for the purpose of organizing their lives.

Chapter Twenty - The War over Reality
Discover May 2017

"The sub title says this: Quantum Physics may be understood but scientists still don't agree what it means."

That pretty much sums it up. Merriam Webster defines "understand" as, "to know how (something) works or happens." So my question is, "If Quantum Physics is understood, how is it that scientists do not agree?" Could it be that they haven't defined their terms unambiguously? Yes, definitely. Perhaps if they illustrated the objects they could arrive at some sort of consensus. Could it be they haven't illustrated their main objects the wave function? Yes, because there are no objects in quantum "physics." That is a misnomer. There is nothing physical about the abstractions and nothing scientific about the probabilistic nature of quantum.

Science is about explaining phenomena using objects, not describing phenomena or proposing objects and then estimating probabilities of their existence. There are only possibilities or impossibilities. We can only determine this when we have defined our Key Terms, illustrated our objects and then made rational assumptions in our hypothesis.

The article begins by asking the question, "What if scientists didn't believe the earth orbits the sun? What if they viewed the heliocentric model of the solar system as just some abstract mathematical tool...?" It goes on to say that it would be preposterous if it was claimed one can not "truly know" whether the earth orbits the sun. Well, except when it comes to quantum mechanics (QM). After all, even if quantum is "baffling" we have to accept it because "experiments have repeatedly proven its weird predictions with phenomenal accuracy." We wouldn't have the amazing technology that drives our world economy and electronics wouldn't work. QM even explains "why the sky is blue!" Yet, in spite of all this, scientists don't agree on "what it means and what it says about the nature of reality." As we learned in the Premier issue of the Rational Scientist, science isn't about belief, truth, knowledge or other types of opinion. Science is about explaining. Hey, I have a suggestion: Why not start by defining reality, what it means to exist? That might help. Just a thought, but, carry on.

We are told that there are at least a dozen radically different takes on QM, and that at least some physicists "deny that quantum mechanics describes any sort of objective reality." There's Many Worlds and QBism, for example, but scientists don't understand what the central equation of QM, the wave function, represents. Wfew! I thought it might just be silly me, relying, as I do, on illustrations of objects to understand how they relate to explain phenomena. Oh well, I guess I'll wait until the movie comes out.

According to Clements University physicist, Antony Valentini, "There's no standard interpretation." That's understandable, since there are no objects, only abstract concepts, or scientific definition of their Key Terms, like exist, it's impossible to have a standard interpretation. You QM folk really need to get together and come up with some illustrations and some definitions.

The author asks the question, "If scientists can't agree on - or don't know – what their reigning theory is all about, does that mean that they have hit a wall in terms of understanding the world?" Well, no, not exactly, Tim. "Agreeing" or "knowing" has nothing to do with science or with understanding, but it does help to have a rational explanation! If scientists can not agree on what their central theory is, then there is no theory. And, without illustrating their objects and defining their key terms there is no hypothesis. Without a hypothesis there can be no theory or understanding possible.

Tim says, "If there's one thing certain about the quantum world, it's that nothing is ever settled." Close, but I'd say, in science, nothing is ever settled. There is only possible or not possible, and something "better" can always come along.

I agree, in part, with Tim when he tells us that the confusion began with Bhor arguing with Einstein and proposing that "physicists must give up the idea of a reality that exists independently of their own measurements." Not just with the atomic world, either. At least there is something that Einstein and I agree on; the moon is there whether or not anyone observes it.

Central to Quantum "theory" is that particles don't have any properties until some lab rat measures them. As Tim points out, "the properties literally come into being at the time of the measurement. I disagree with the author when he answers his own question, "What makes quantum mechanics so confounding?" It's not that "photons behave as particles with detectors present and like waves without detectors." It's that science uses experiments at all to "explain" anything about particle physics. To these guys it's a forgone conclusion that there are particles when they have no justifiable way of explaining certain aspects of reality like the force of "pull" with particles. Besides, experiments never explain anything to begin with. Experiments only ever convince the experimenters and gullible public that something has been "proven."

Our author, Mr. Tim Folgers, asks a very good question here: "Are these 'quantum waves' as real as the earth beneath your feet, or are they purely mathematical constructs without any physical existence?" The answer would be evident had terms been defined and objects illustrated. Let's see if we can answer his question by defining some terms.

"Quantum: a discrete quantity of energy proportional in magnitude to the frequency of the radiation it represents." – Oxford Dictionaries.com

Cambridge says about wave, "to raise your hand and move it from side to side as a way of greeting someone". Nope, surely not that. How about this one from dictionary dot com: "In physics, any regularly recurring event, such as surf coming in toward a beach, that can be thought of as a disturbance moving through a medium." That makes more sense in this context, but Physics classroom dot com says this about waves: *"waves* involve the transport of energy without the transport of matter."

A hand waving and the surf coming in towards the beach make sense at least! An object, hand, is moving. Ocean water is moving. How does a discrete quantity or disturbance move? This is reification at its best! We can only apply verbs to nouns, yet moving motion, or transporting an ability is typical of mathematical theorists.

The answer to Tim's question is: NO, quantum waves are not "real as the earth beneath your feet." Real is a synonym for exist. Exist is the one term that makes or breaks QM propositions and should be THE most important term in all of physics.

Exist; object with location; something somewhere. An object is that which has shape. An object exists if it is physically present, that is, it has location with respect to all other objects.

The mechanic turns concepts into objects and verbs into the nouns of reality. They all need to go back to their early readers. See spot run! The particle physicist will have run there before Spot arrives on the scene and still there after Spot is long gone. In the case of the quantum magician, they will measure run and then Spot will pop into existence.

Everett's Many Worlds theory says that all events happen simultaneously in parallel universes. Leave it to David Wallace, philosopher of physics, to tell us, "It leaves the physics unchanged, and holds onto the idea that science is supposed to give us a description of what is going on, even if what's going on is weirder than we thought."

Sorry to disagree with you, David, but science is supposed to explain what we propose, not describe what we see. If we ask

"Why the sky is blue?" the answer, "Because we pointed at the color and named it blue" doesn't tell us anything about how blue light is scattering differently than red in the atmosphere. The answer "we named it blue" is not scientific because it explains nothing about the phenomenon. But if we propose light scattering using photons and Snell's Law, we better be able to back that up with illustrations and reason. If we are talking about the double slit experiment, and the claim is that photons leave as particles, travel as waves and arrive as particles, then we had damn well better be able to provide a series of illustrations or a movie showing us how these little shape shifters perform this magical act.

Other physicists hold that the wave function doesn't really represent any physical entity, it only expresses the probability that a particular result may follow from our experiment. Erwin Shrodinger proposes his cat in a box thought experiment to underscore the absurdity of Bhor's idea of measurement creating reality. In other words, the wave function can not be real. The cat is dead or alive, not both; despite any probabilistic outcome we give it. We only know the result after opening the box. Referring to the double slit experiment, the author asks the question, "If the wave function isn't real, what creates those light and dark bands?" The answer is simple: diffraction. These same results will be observed by using a laser and a needle as in Gaede's demonstration or Newton's hair experiment. This has nothing to do with a particle either going through one slit or the other as there are no slits to choose from!

The PBR theorem argues that the wave function must be real as there is no way that one physical state can "be compatible with so many different wave functions." The rationale is that the wave function must correspond to something that is real "proving that the wave function actually exists…" Circular reasoning much?

QBism combines QM with Bayesian probability which revises the odds of something occurring as more is learned about a given event. QBist Schack says "If QBism says one radical and important thing about the nature of reality, then observer participancy is it." Once again the observer is put in the center of the method of inquiry. Rational science "kills the observer" removing subjectivity. How else can something be objective if not by removing subjectivity? The QBist just adds more and more experiments, and more and more subjectivity. QBism says that there is no experiment separate from the experimentalist. Well, that does make sense, but experiments are extra-scientific, and what one does AFTER the conference, on his own time.

Since Schack sees Quantum mechanics as a "set of rules" for conducting experiments, naturally he says. "Whether you see a wave or a particle depends on what question you ask." Even more revealing is this statement, "Quantum mechanics is a useful guide to action: it tells you how to put together our experimental apparatus so that it works in the end." Yep! This is why we don't do experiments in science. It only confirms in the mind of the experimentalist that he has "proven" something.

So, the article asks, "What is the quantum world made of? Particles? Waves? We get our answer from Valintini who defers to DeBroglie's Pilot Wave theory. Both particles and waves are real, since a particle rides on a wave. Pilot waves exist in multiple dimensions and they are real physical entities. A particle may choose to go through one slit or the other, but its pilot wave goes through both at the same time! Changing the number of detectors changes the Pilot Wave. Valintini's revived Pilot Theory may be vindicated because he has found a way to test it experimentally in spite of the fact that none of the other interpreters of quantum can make that claim. Some how I knew that was coming. It seems that Pilot Wave theory predicts certain things about the Big Bang that we will be able to detect in small anomalies in the Cosmic Microwave Background.

Of course Big Bang Theory and CMB play into this.

The article ends on a very positive note when Valintini admits, "...really if we are going to be honest, as scientists, if a member of the public asks us what is the meaning of our most basic theory of physics, I think all we have to say is we don't know."

Chapter Twenty One - The Neutrino Puzzle
Scientific American, October 2018

The author takes us inside a cave and introduces us to a huge apparatus which is part of Fermilab and known as the NuMI Off-Axis Electron Neutrino Appearance experiment. About a thousand miles away, an even larger device known as NOVA will detect neutrinos as they pass through NuMI, the earth between Illinois and Minnesota and the NOVA detector.

The idea of the experiment is to view the light trails left in the wake of neutrinos traveling at near light speed as they pass through, well, everything that is in their way. Because neutrinos are nearly massless and moving so fast, these "ghostly particles" can even pass through the "empty space" between atoms of our bodies at the rate of trillions per hour "without a trace." None-the-less several times per day a neutrino will interact with the detector knocking loose particles that leave light trails scientists can see. NOVA catches these fastballs and claims to have been doing this since 2014. This is the "longest distance" experiment ever devised, but if you think this one is big, wait until you see DUNE. The Deep Underground Neutrino Experiment will use the accelerator at Fermilab in Illinois to smash graphite producing a beam of neutrinos that will be fired at a detector buried deep in a gold mine in South Dakota 500km further away.

The name DUNE is appropriate because, like Frank Herbert's book, this is the stuff of science fantasy, not science! And their experiment will come up dry as a desert planet when it comes to neutrinos, because like ghosts, they do not exist. The NuMI and DUNE projects will use up more gold than was found in that mine. Like the Superconducting Super Collider fiasco, the projects are doomed to fail so they should leave NuMI buried below the earth, and abandon DUNE like the miners abandoned the old Homestake gold mine it is in years ago. Without over runs, DUNE will cost 1.5 billion dollars.

Researchers from all over the world are involved and even CERN is investing in the project. A spokesperson for DUNE said, "We hope DUNE will do for neutrinos what the LHC did for Higgs." Oh, it will.

They will claim success even though nothing will be discovered and nothing accomplished to further our understanding of Father Universe. Folks will sashay down isles to receive pats on the back, bouquets of roses, tiaras and Nobel prizes. Billions will be spent, careers will be advanced and particle physics will march on and on and on. These folks are excited because the neutrino is the first particle to "break from the so-called Standard Model, physicists' best description of nature's fundamental particles and the rules that govern them."

In other words, they can create more and more "particles" to add to the ever growing list of reifications currently "predicted" by mathematical theorists, and they get more theories, ad hoc, to justify their previous calculations. But there are no discrete particles to explain the force of pull (hence the fantastic Higgs field contrivance), and, as we learned in the previous issue of The Rational Scientist, science doesn't predict. At best, science can only explain consummated events. The list of particles will continue to grow, because theorists are turning abstract mathematical concepts into physical objects via reification. Mass, fields and charge are what objects are doing, not objects themselves.

So what are neutrinos? Neutrinos were thought to be massless, charge-less particles that travel near the speed of light and carry away energy and momentum from radioactive particles that are going through the process of Beta-decay. However, earlier experiments were unable to detect the number of solar neutrinos that mathematicians expected, so once again a theory needed an ad hoc revision to make the numbers work. All would be fine if neutrinos actually had some tiny amount of mass. Particle physics starts with irrational assumptions in their hypothesis and then builds a theory on it and hypothesis upon that and a theory upon that until there is no hint of reality left.

Of course any so-called particle should have mass (a property of a physical body that we measure in kilograms), because without mass we are stuck with yet another zero dimensional particle. Rational scientists understand that in order to exist, an object must have length, width and height; the three mutually orthogonal directions an object can face. To exist, an object must have shape and location with respect to all other objects.

So now that physicists have made neutrinos more interesting (though not more plausible) with an ever so slight mass, some phyz whiz or group will make their bones with this, rest assured. They have something that requires a "new physics" allowing particles to "acquire mass" from something other than the Higgs field, giving it a lot of appeal. New particles and new fields; researchers are having a field day! Since the LHC has not produced a new particle, that is, one NOT predicted by the Standard Model, such as Dark Matter particles, interest has turned elsewhere. Fermilab's neutrino expert, Stephen Parke says, "Some people are betting on the LHC with their careers. Others of us are betting on neutrinos."

In the world of probabilities nothing beats a warm bowl of Gambles Quark Noodle Soup!

We understand that the theoretical scientist is a gambler since they are into probabilities rather than possibilities.

Particles will always breed more particles and the more "answers" the more questions will arise. Parke says, "The thing about neutrinos is, the more you understand, the more questions you have." Well, that should keep them busy for another 100 years.

According to the interview with Parke, DUNE will focus on neutrino oscillations allowing these nudnicks to determine the identities of various flavors of neutrinos. Apparently these neutrinos can change from one to the other so that a muon neutrino leaving FERMILAB will arrive at DUNE's far detector as an electron neutrino. The only way that this can happen is if neutrinos have some amount of mass. Erm, uh.. the only way anything can DO anything is if it has some amount of mass, but let's continue shall we? As neutrinos travel through space, the differing amounts of mass cause them to move at different velocities in accordance with Einstein's relativity. Oh good, I was worried there for a minute. We already have particles NOT predicted by the Standard Model. By counting the numbers of neutrinos that leave as one flavor and arrive as another flavor, DUNE will "determine how the different neutrino masses compare to each other." They want to find out if there are two heavyweights and one light weight or just the opposite. "By firing neutrinos through matter, you can determine the difference very easily, and the further the distance the clearer the signal," says Thomson, "That's a bit of physics that DUNE is absolutely guaranteed to nail within a few years."

Let's see if we can understand what these folks are blathering on about, shall we? "The" neutrino is a contrivance of Wolfgang Pauli to "explain" away the failure of a Standard Model "prediction." Comparing the energy and angular momentum of pre and post Beta decay particles it was found that conservation did not hold. Pauli proposed that yet another elementary particle must be responsible for carrying off the missing values of energy and momentum.

Pauli's zero mass neutrino has been replaced with three non-zero mass neutrinos: Electron, Muon and Tau neutrinos.

Using the analogy of polarizing lenses, the particle phiz tell us that the three different flavors, or types of neutrino are interacting with each other mid flight and changing back and forth (those shape shifting little demons) into each other. They call this shape shifting "neutrino oscillations."

Particles sometimes act as waves, their frequencies being proportional to their energies. These "matter waves" interact constructively or destructively with each other. If we are talking about two waves of identical frequency: when they interfere constructively (peaks line up) their constructive interference produces larger peaks. When interacting destructively (the peak of one lines up with the trough of another) their destructive interference amounts to smaller peaks.

If we are talking about two waves of different frequencies, the resulting wave is not a similar wave of larger or smaller amplitude, and there is no clear wavelength. There will be points where the waves interfere constructively and where they interfere destructively. At points where one wave is crossing the zero line and the other is at a maximum or minimum point, the composite waveform will appear to be like the first wave. Reverse this, and the composite wave appears to be like the second wave. The rate, or frequency that this process repeats is mathematically the difference between the two waves.

So, what we have here is an analogy of how one neutrino can shape shift into another. To add irrationality to the ridiculous, they are not talking about different neutrinos of different flavors (masses) they are saying that actually, the neutrino is interfering with itself! They gave neutrinos mass because without shape shifting they could not explain missing muon neutrinos and solar electron neutrinos. Experiments don't practically detect tau neutrinos, so if atmospheric muon neutrinos become tau neutrinos and electron neutrinos become muon or tau neutrinos, that would explain why the predicted numbers are not being experimentally detected.

Neutrino oscillations are named such because it's not a one way street. One type will change back and forth to another type as it travels along. Quantum magic's probabilistic nature allows that any given neutrino will change after a certain distance and will become a blend of two or even all three types.

So what really are we talking about when we say constructive or destructive interference? We aren't saying there are two matter waves, we are saying that one acts as two and that a single neutrino is interfering with itself! In other words, neutrinos have no definite mass. When an electron neutrino is created it is in a quantum state of two different masses...er..in... one. A muon neutrino also has two masses. The two masses interfere with each other as the neutrino propagates. The difference in the two frequencies will combine in such a way that sometimes the "mixed" wave is mostly like one mass and frequency and at other times like the other.

As crests and troughs (well really other components like mass) add and subtract from each other, sometimes we have a muon and sometimes we have an electron neutrino "matter wave." AND sometimes the neutrino is neither!

Until the neutrino, particles like electrons and protons "acquired" their mass from interacting with the Higgs Field. Of course, how a particle can have surface to surface contact with a field, is anyone's guess. Wait, it can't! Objects never interact with concepts, silly! Oh well, let's humor them and trudge on. It seems that the Higgs magic only works with particles that are ambidextrous. But neutrinos only spin in one direction relative to their direction of travel. In other words, they are all lefties, so they must get their mass from some, as yet unknown, other field. We predicted that was coming. According to the director of Fermilab, "There's probably a completely different mechanism, and therefore there should be other particles associated with how that happens."

Something else that gets these guys excited is that neutrinos might be what are called Majorana particles. A Majorana particle is its own anti- particle. Standard Model predicts that particles come in pairs, so that an electron has an anti-electron, or positron, for example. They are twins, but one is left-handed and the other is right handed. One carries charge and the other does not. Since neutrinos are too lazy to carry charge, and that is what distinguishes one from the other, Majorana particles must get their mass from a source other than Higgs. The math tells them that there must be a heavy neutrino, undiscovered thus far, that weighs a trillion times more than anything else they know about.

DUNE won't "directly test" whether neutrinos are Majorana particles, of course, but the results of their experiments will help them analyze results being obtained around the world. Yep, we predicted that, too. None-the-less, all these experiments may provide such answers as, "Why is the Universe made up of matter instead of anti-matter?" After all, cosmo scientists say that after the Big Bang that both should be equally abundant.

Neutrinos may hold the answer to this mystery and DUNE might find antimatter muon neutrinos turning into electron neutrinos at a rate indicative of something or other. It may be that the only difference between neutrinos and anti-neutrinos would be their handedness. It would make particle physicists happy to know that there are an equal number of lefties as righties. DUNE could also tell researchers if there are only three flavors, or if it is like Baskin Robbins 32 flavors of ice cream. That would really, really make particle physicists happy!

The article ends on a good note, in my opinion, when it says, "To succeed, DUNE will have to overcome the political and funding hurdles that have killed large physics projects before."

We can only hope.

Chapter Twenty Two - Quantum Computing

The Quantum magicians are so far out there that there is no distinguishing their theories from the supernatural or science fiction type theories rampant on the interwebs. A case in point: My tattooist was very adamant that quantum computers have been created and that the D-wave computer had been used by NASA, Google and the scientists at CERN, inadvertently creating a warp in spacetime, altering forever our particular time line. He had dozens of examples of what has been termed the Mandela Effect, showing how what you remember about certain scenes in Forrest Gump and Star Wars, for example, has been changed. These, we are told, are NOT false memories. Instead, they are effects of our altered timeline due to use of the D-wave quantum computer. This is the danger of quantum magic. Not warped time lines, but that people can believe in such irrationality!

The next day, I heard someone listening to a recording of a church service. The pastor said that quantum teaches us that "we can change our reality," and that is "the power of prayer." No need for a computer at all!

There is no difference between Quantum Magic and religion. Both are irrational explanations for reality. Let's see what wikedpedia and other web based resources have to say about quantum computing.

Quantum Computing

"Quantum computing is computing using quantum-mechanical phenomena, such as superposition and entanglement." - Wiki

IBM is working on Quantum computer technology, what do they say?

"Quantum computers leverage different physical phenomena — superposition, entanglement, and interference — to manipulate information. To do this, we rely on different physical devices: quantum bits, or qubits. ... Entangled particles behave together as

a system in ways that cannot be explained using classical logic."
https://www.research.ibm.com/ibm-q/learn/what-is-quantum-computing

Well, then, let's define those terms and then see what they are all about before applying that to computing.

Superposition
"A quantum computer ... maintains a sequence of qubits, which can represent a one, a zero, or any quantum superposition of those two qubit states; a pair of qubits can be in any quantum superposition of 4 states, and three qubits in any superposition of 8 states." – WIKI

Entanglement
"Quantum entanglement is a physical phenomenon which occurs when pairs or groups of particles are generated, interact, or share spatial proximity in ways such that the quantum state of each particle cannot be described independently of the state of the other(s), even when the particles are separated by a large distance—instead, a quantum state must be described for the system as a whole." – WIKI

Interference
"Physics (of light or other electromagnetic wave forms) interact to produce interference. ... the combination *of* two *or* more electromagnetic waveforms to form *a* resultant wave in which the displacement is either reinforced *or* cancelled." Oxford Dictionary of English

What is entanglement?

In April of 2016, Frank Wilczek, Nobel Prize recipient and physicist at the Massachusetts Institute of Technology, was the month's featured columnist at Quantized, an on-line magazine.

In order to understand entanglement, Frank uses the more common object, cakes, to "pry the subtlety of entanglement itself apart from the general oddity of quantum theory."

We are told that entanglement arises from our having a partial knowledge "of the state of two systems." Hmmm…a not so subtle change from an object to a system. Looks like bait and switch to me. So we have two cakes one square and one round giving us four possibilities: two squares, two circles and either one square and one round or one round and one square, depending on your point of view. If the shape of one doesn't tell us anything about the shape of the other then they are said to be independent, but if they are entangled, information about one will give us "knowledge of the other."

In "extreme entanglement," if one is circular we know the other is also circular. This lack of independence is "essentially the same" in quantum theory of entanglement. Of course, the difference is that we aren't talking about cakes we are talking about mathematical "objects" called wave functions. Ah, now I see why the bait and switch before. We aren't really talking about objects, like cakes, or as in something physically present. We are talking about mathematical abstractions.

Entanglement doesn't really happen with cakes, but entanglement does occur naturally, like when CERN collides particles in their accelerator. The exception is NO entanglement. Next it gets a bit complicated when Frank gives us an example of a molecule being a composite system whose lowest energy state is a "highly entangled state of its electrons and nuclei." The positions of these nuclei are not independent because when protons move the electrons must also move. He is talking about Heisenberg's Uncertainty Principle, where one can only know the position or the speed of travel and not both at the same time.

One way to create an entangled state is to obtain only partial information about a system by measurement. We can learn, for example, that they have the same shape but not what that shape is.

Enter complementarity. We can add the property of color to our shapes and have, perhaps four states, a red and blue circle; a red and blue square. With complementarity our cake doesn't have to have BOTH color and shape. Imagine that for a moment, a red cake with no shape. You can't can you? Only an object, "that which has shape" can have color. There is no stand alone object called "red." Unfortunately for common sense, experiments say otherwise. When they measure shape of their quantum cake, they loose information about its color.

What this tells quantum magicians is two things: If you can't measure a system, it doesn't "need to exist" and when you measure it, it changes the measured system. What these folks actually measure in their experiments, is angular momentum of electrons. But back to Frank's delicious cakes. Oh wait, that's not a property yet. If you measure the color or shape of one of the pair, its twin is, "equally likely to be red or blue", "equally likely to be square or circle."

If you measure both of the entangled pair for either color or shape, they are always the same. If instead you measure shape of one and then the color of the other there can be any state. In other words, if one is square the other is "equally likely " to be either red or blue. And, gasp, this happens regardless of the distance the complimentary two are apart from each other. Measuring one instantly affects the other. This is Immediate Action at A Distance. Information transmitted at a speed faster than light!

"But does it? Until I *know* the result you obtained, I don't know what to expect. I gain useful information when I learn the result you've measured, not at the moment you measure it. And any message revealing the result you measured must be transmitted in some concrete physical way, slower (presumably) than the speed of light."

The bottom line is that it is impossible to have independent systems in quantum. Frank goes on to describe entangled histories where it is impossible to "assign a definite state to our system at each moment in time." In other words, if they measure at two different times it can be either square or circle "but that our observations leave both alternatives in play." Quantum cakes, then have no determinate shape or color until they are observed.

"A definite state can branch into mutually contradictory historical trajectories that later come together."

"Our q-on (ED cake) might be prepared in the red state at an earlier time, and measured to be in the blue state at a subsequent time. "

What is superposition?

Let's see if we can find out from a May, 2018 Scientific American article entitled, "Quantum Physics May Be Even Spookier than You Think."

A superposition is when a particle "seems to be in two or more places or states at once." The article looks at a proposed experiment which reveals "surprising hidden mechanics."

The experiment proposes to look at a photon to see "where it resides" when in superposition. The two researches say that it will be "even stranger" than the idea of being two places at once.

The article discusses the classical two slit experiment where a laser is fired at to parallel slits to observe photons acting like both a particle and a wave.

Firing a laser at two slits produces a fringe pattern on the wall. The experiment is interpreted to mean that photons passing through one slit are interfering with those passing through the other. What the article's author finds "odd" is that this happens when only one photon is fired at a time. So, in other words, a single particle is passing through both slits at the same time – superposition! Sorry, it's not odd, it's impossible. There are no single photon producing emitters, but if there were, a photon can not be two places at once. What is happening is simple light diffraction. Newton's hair, Young wire and Gaede's needle experiment will show you this. Fire a laser pointer at the wall. Place a needle in a cork between the laser and the wall. You will observe fringe patterns or interference wave patterns. Nothing odd or impossible about this. We cover this in detail in: Rational Science Vol. I; Chapter Eleven – Light, Particle or Wave?, Chapter Twelve – Smoke and Half Silvered Mirrors, Chapter Twenty – Quantum Magic

Of course when you measure which slit a "particle" goes through the "superposition collapses." Yep, that's quantum. When you measure a particle the waveform collapses.

Therein lies the quandary. Lab rats need to observe this to see what is actually happening but when they look at it goes poof!

The brilliant researchers think they have found a way to fool Mother Nature, by deducing "something about quantum particles BEFORE measuring them." Uh, yeah, it's called a hypothesis and theory. Ever heard of it? First you visualize the object (hypothesis) and then explain how it mediates the phenomenon (theory). Anyways, these nudnicks have a different approach they call the Two-state-vector formalism (TSVF). It should work because, "quantum events are in some sense determined by quantum states not just in the past—but also in the future. That is, the TSVF assumes quantum mechanics works the same way both forward and backward in time. From this perspective, causes can seem to propagate backward in time, occurring *after* their effects."

Actually what they do is "gain knowledge retrospectively" by choosing where to look for a particle beforehand rather than measuring it after it gets there. "This is called post-selection, and it supplies more information than any unconditional peek at outcomes ever could. This is because the particle's state at any instant is being evaluated retrospectively in light of its entire history, up to and including measurement." For Sagan's sake people, this is what science is supposed to do. One looks at a consummated event and then explains what happened. NOT: Hey look, I pointed the laser at the wall and there is a red dot there right where I predicted it would be! Good thing I had my meter set up over there to measure it.

The researchers have joined with other quantum magicians and have concocted an even more elaborate experiment using shutter photons and three mirrors. By observing "self-interference" they can conclude, after the fact, the particle was in one of three boxes at a particular time. If the photon is in more than one box at the same time, now here, then there, it would illustrate "nonlocal existence of quantum particles."

Not surprisingly, I can't find anything about the results of Avshalom Elitzur's TSVF experiment (at the Perimeter Institute). Guess he couldn't pull it off, or couldn't get the lab time.

OK, now that we know a little about superposition, entanglement and interference, let's see how those might apply to…

Quantum Computers

Let's see what the head of Intel's Quantum Computing told Scientific America's Larry Greenmeir in May of 2018. From his article:

How Close Are We—Really—to Building a Quantum Computer?

The discussion is about "a technology that doesn't exist yet" …"the world's first meaningful quantum computer." Even though the bigs IBM, Microsoft, Intel and Google are working on it is "most likely a decade away."

They are busy packing more and more qubits onto computer chips and trying to solve cooling issues and inventing new technologies to handle the speed and processing power.

Unlike regular computers, a quantum computer process qubits which can be ether 0, or 1 or both. 0s and 1s are dependent on the spin of an electron. Clockwise could be a one and counter clockwise a zero. With superposition, the qubit would be in both states, so spinning in both directions at once, I suppose. Qubits are very unstable and need to be cooled to 250 times colder than deep space, that is, 20 milikelvins. I think that the researchers are a bit unstable and probably should be put in the cooler for a while themselves.

Intel showcased their Tangle Lake 49 cubit processor at the U.S. Consumer Electronics Show earlier in 2018, and in my home town at the University of Texas at Austin, when it is not being used to track storms, the super computer named Stampede is used to simulate "up to" a 42 cubit processor

Jim Clark, Director of quantum hardware at Intel Labs, explained qubits using the analogy of a coin. A regular computer's transistor indicates the coin as either heads or tails. But you can think of quantum computing like this: While the coin is spinning the qubit can represent both heads and tails. "The ability to sample a large number of possible combinations" is called the state space. Tossing two coins at once gives four possible states, three coins gives us 8, and 50 coins would represent more states than the fastest super computer can compute. Finally, three hundred coins would represent more states than atoms in existence.

Qubits, however, stop spinning and "collapse into a particular state" but magicians need the spinning to continue for a very long time. Small disturbances like temperature fluctuations, vibrations, or a technician blowing an ass trumpet could cause spinning to stop along with a resulting loss of data. To make qubits more stable they are cooled down to just above absolute zero using an isotope of helium.

Intel, IBM and Google's quantum computers are called superconducting systems and all require cooling, but there are different types of qubits, or ways to manipulate qubits. Other methods use oscillating ions held in a vacuum by lasers, and electrons whose spins are controlled by microwave pulses. The later probably more commercially viable.

Intel runs simulations on a supercomputer and finds that it takes around five trillion transistors to simulate 42 qubits, but likely a useful system will require at least a million qubits. What kind of software is needed can't be determined until they have a computer with a few hundred or a few thousand qubits.

A superconducting computer would be prohibitively large, but spin cubits are a million times smaller and the more likely option. Possible future uses would be artificial intelligence or security (cryptography).

Clark projects at least ten years into the future before quantum computer system approaching a few thousand qubits will be in existence.

In Wired's article of 16 Feb 2018, entitled, "What is quantum computing?"

We are told this:

"Quantum computing takes advantage of the strange ability of subatomic particles to exist in more than one state at any time." And "A qubit can be thought of like an imaginary sphere."

Now, I say that a pie may be square or it may be round, but it can NOT be both square and round at the same time. The pie can be both round and red, but whatever, a qubit is doing, it is not a 0, a 1 and a 0/1 at the same time. Whatever results the quantum computer is telling researchers, rest assured it is not because of superposition. If these persons wish to play with their imaginary playmates, fine, but real objects do not behave that way.

"How far away are quantum computers?"
IBM has a 5-qubit and a 20-qubit system

"D-Wave, a company based in Vancouver, Canada, has already created what it is calling a 2,000-qubit system although many researchers don't consider the D-wave systems to be true quantum computers."

Intel recently announced that it has a way to make quantum chips from silicon.

What can quantum computers do that normal ones can't?

They can't do anything since they don't really exist. BUT let's humor them.

"In July 2016, Google engineers used a quantum device to simulate a hydrogen molecule for the first time, and since them

IBM has managed to model the behaviour of even more complex molecules."

According to Engadget, Technology Review and other sources, "Scientists at IBM were able to simulate beryllium hydride (BeH2) on a seven-qubit quantum processor last September..." So, they claim they were able to model a molecule in a few minutes as opposed to five days. BUT this was using a quantum processor and a super computer, not a quantum computer.

As we mentioned earlier, D-wave sells what they are disputably calling a quantum computer. From their website:

"D-Wave's flagship product, the 2000 qubit D-Wave 2000Q quantum computer, is the most advanced quantum computer in the world. It is based on a novel type of superconducting processor that uses quantum mechanics to massively accelerate computation."

However, many dispute that the machine has shown any quantum speed up at all, and that it is not even clear if superposition is actually taking place. Likely, the effects are due to non-quantum thermal annealing effects. I'd say it is more than likely. We can end the debate right here and right now.

Superposition is impossible, therefore quantum computers can not possibly work as claimed using superposition or any other kind of quantum magic.

Rest assured Ray, either false memories, or altered film is responsible for the "changes in our timeline." Even if quantum computers existed, and even if superposition were possible, time and space are concepts. We can't bend, warp or alter, in any way, concepts.

Chapter Twenty Three - Math is Descriptive

We have learned that math is descriptive and so naturally mathemagical physicists such as Symanek want us to believe that "Science doesn't explain. Science describes."

This way, folks can describe some abstract mathematical concept, such as black holes, prove it with their math, and then tell you that their irrational physical interpretation was predicted and therefore we can assume, correct. Many are banking on the fact that you won't understand their math and so, even if there are mathematical errors, you will never see them. They will tell you that you need to go to a university and get a degree in mathematical physics.

When an extraordinary mathematician like Stephen J. Crothers comes along and destroys these mathemagical proofs, you'll never know it. You won't understand the math, and folks protecting their defunct theories will, at best, marginalize their detractors.

We don't need to understand math, any more than we need to be an ichthyologist to know when a fish stinks. We need look no further than the irrational physical interpretations. On the other hand, if mathematical theories are purely descriptive, then the physical interpretation is up for grabs, that is, the beauty is in the eye of the beholder.

We saw that Einstein's field equation, while elegant and beautiful to the relativist, can not possibly be proof of warped spacetime. Regardless of how error free an equation is, if it predicts the impossible, it is time to erase the whiteboard and start over.

Perhaps one of the very last equations, whose physical interpretation *is* possible, is a derivative of Maxwell's equations for electromagnetism: the wavelength/frequency equation:

$$\lambda = fc$$

Maxwell never understood the underlying physical mechanism behind electromagnetism, but, like Faraday before him, did understand that there must be some"thing" there: matter.

He referred to field as, "the space in the neighborhood of the electric and magnetic bodies… in that space there is matter in motion" - A dynamical theory of the electromagnetic field , Phil. Trans. 155 (1865) 459 – 512

"I cannot conceive curved lines of force without the conditions of a physical existence in that intermediate space." Michael Faraday, On the Physical Character of the Lines of Magnetic Force, Philosophical Magazine 3 (4), (June 1852) in Experimental researches in electricity, Vol. 3, Bernard Quaritch, London (1855) 407 – 437

Though they never understood electricity and magnetism in terms of what it actually is, the equations are still very useful for designing optics, radio equipment, radar, and many other devices. This is engineering, which is very useful for building (technology), but fields and lines of force explain (science) nothing.

We can apply the equations and their derivatives to many things, and we can also ascribe many underlying physical mechanisms claiming that they are the mediators of the phenomena.

As Bill Gaede, originator of the Rope Hypothesis, says, "Maxwell's equation for wavelength is the equation of the rope."

Here we have a simple physical entity: a two strand rope comprised of an electric thread and a magnetic thread. Let's apply that to the phenomenon light. Frequency and wavelength are inversely proportional and the speed of light, c, remains constant. The two threads are wrapped helically around a common axis and "travel" at ninety degrees to each other.

Take a two strand rope of a particular length and twist it. The more you twist it, the more links per unit length are seen. A demonstration of inverse proportionality. Tie one end to a pole and place a clothespin on it near the pole. Twist the rope from the other end and see how quickly the clothespin responds by moving. Very, very fast! Immediately, it seems, without a time lapse camera.

Among many other issues, particle physics has problems with wavelength and atomic size. Ask this question in Quora, or other science forums:

If light waves are larger than an atom, how can the electrons in an atom absorb light?

And get this:

"Wavelength isn't important for absorption. The energy is what matters, and that the wave packet of each photon is large enough to overlap with the atom at some instant in time."

The person goes on to say that by the time the photon arrives at the next atom "the wave function has vanished."

How is it possible that the wavelength of an electron is much larger than the size of an atom?

"Size of an atom is 10 power -15m. Electron wavelength specifies its energy. Energy and size are two different entities."

How does a single atom absorb a photon when an atom is smaller than the photon's wavelength?

"The photon that might be absorbed is not precisely localized at the scale of a single atom until it is absorbed. IOW, there is some prior probability that many different atoms of the same element in a sample could absorb a particular photon, but at most one of these atoms will absorb the photon. Quantum mechanics only provides the probabilities. We can't say for sure in advance which atom will absorb the photon."

So, we see that answers to the question of atom size in regards to wavelength include irrational statements such as:

1) "It is energy, not wavelength that is important"

2) wavepackets must overlap

3) wavefunction vanishes (collapses)

4) It's an issue of probabilities, and a particular atom doesn't have location until it is absorbed

Of course we need objects to do physics. Energy, wavepackets, wavefunctions and probabilities have nothing to do with rational science, or how Mother Nature runs her shop.

How does the Rope Model provide an answer to this question?

Wavelength IS important, as wavelength is the inverse of frequency. What is considered a photon is the torsion of a single link as it separates at the atom. A low frequency will have a longer wavelength, but what is "seen" at the atom is a single link at a time. So, how does a tiny nanometer sized atom swallow up a 60 mile long radio wave? It doesn't! At the "sending" atom and the "receiving" atom the links are the size of the atom, it is at the middle, between the two atoms where the wavelength is longest. Tie a rope to a chair leg or tree at ground level. Whip the rope. Observe the link as it "travels" along the length. Whip it fast and see smaller links. But whether slow or fast, the link is always longer at the middle point.

Keep in mind that the atom is like an anchor point, the rope is straight, torsion is bi-directional, and the rope is unraveling at the atoms.

The energy of a photon is derived by taking the difference between the ground state and the excited state. The transition is this distance (or pump height). Using the transition and lambda along with Rydberg's constant we can derive Rydberg's equation for spectral emission.

We can progress from this to other equations that depend on the Planck constant, such as dealing with the photoelectric effect, Eistein's field equations and more. By applying the rope model, we can propose a physical interpretation.

Radio waves have energy on the order of a femto electron volt and can have wavelengths of 1000's of kilometers. Wavelengths can vary from a Planck length to the distance across "the universe."

Maxwell's formula can be reordered like this:

E=barHc/lambda

Where:
E=energy
BarH is Planck's constant, or, 6.626×10^{-34}

What does all of this mean, if anything, considering the rope model for light?

The rope architecture insures a limited number of phenomena. Motion of thread and rope manifests as: torsion, tension, vibration, friction, rotation and pumping. To the degree that equations represent these phenomena is the degree that they represent the rope.

We do not have to make the rope fit the mathmagician's equations, as we can see from Einstein's field equation, for example. There is NO rational physical interpretation possible from that mess.

When we hear this: The higher the frequency, the greater the transition. The greater the transition, the higher the energy level.

What does this mean in terms of the rope architecture? Obviously, the transition is related to the rate of the pumping E shell, which is related to the height or distance between the expanded and contracted position of the E shell. The higher the frequency, the faster the atom pumps and the more links per unit length along the rope.

When we hear this: Energy can only be emitted or absorbed in packets this size: $E_{photon} = \bar{h}f$

We can ask what is Planck's constant? We might rationalize that this constant applies because there is a finite amount of thread, and so there is a finite number or ropes. IOW, number of ropes (total combined length) divided by the number of links results in an average link length with an energy equal to 6.626×10^{-34}.

When we hear this: Energy of a photon is equal to the number of photons with a given frequency. We may be able to justify this by saying that Energy is equal to pump speed times friction. Therefore, c is constant because it is the average of friction times pump speed. It is an average because there is a finite amount of thread and every atom is connected to every other atom with a "layer" in every shell and a thread in every proton. The photon is not a particle. It is the torsion "felt" at the surface of an atom as a result of friction and pump speed.

Light is torsion, which is directly proportional to pump speed and friction

And why is it that a smaller wavelength gamma ray has a higher energy than a larger wavelength? Because there are more links per unit length, and hence more torsion relayed between atoms in a given time of measurement.

If the photon isn't a corpuscle, then how is it possible to get images of it?

It's not possible.

All images are computer generated based on data gathered. Producing holographic images of probability amplitudes and phase variations is taking this to an even higher level of abstraction. A holographic image is not an object and probability amplitudes are not atoms.

But even by their own reckoning, electrons are "probability distributions."

Further more, single photon avalanche diodes and single photon detectors are non-existent. Read the tech sheets and find there are no single photon emitters or detectors of any kind. These produce or detect "photon streams" or "average number of photons."

It seems manufacturers cater to the Quantum magician by using their own poetic license.

"Insofar as mathematics describes a rope it applies to reality, and insofar as mathematics applies to reality, it describes the rope." - Albert Einstein

Of course, Einstein was wrong about that. He got his abstract notions confused with the underlying physical mechanisms. His verbs confused with his nouns. Rope and thread are the mediators of electricity, magnetism, light and gravity.

We assume that the mediator of electromagnetism and gravity is the rope. The rope architecture assumes that everything is interconnected, and this insures that every phenomenon is related to every other. Ergo, both gravity and electrostatic "force" are stronger or weaker based on "the inverse of the distance squared.":

$Fg = G.M1.M2/D^2$
$Fe = K.Q1.Q2/D^2$

We also find that c and G are constants, because there is a finite and constant amount of thread and rope, and because ropes remain straight and taut between H atoms..

You may not like the rope model, but it is at least a physical mechanism that can be illustrated. It is visualizable whereas fields and lines of force are not possible or imaginable. An area around an object with points of varying values can only be illustrated with numbers, lines and arrows. We have learned nothing about what the invisible mediator looks like. We do understand that numbers lines and vectors, are concepts, not objects.

What we must do, when it comes to the invisible, is justify our mechanism by illustrating it. We must make the invisible visible. If we can not do this, then we do not understand the phenomenon. The Golden Rule of physics is that we must have objects! Objects that have a physical presence with respect to everything else in existence.

A physical mechanism, that is, objects mediating phenomena, beats abstract concepts any day of the week.

We find that the rope model provides an underlying physical mechanism which mediates all phenomena from electricity, and magnetism, to light and gravity. With it we can explain electrical current, resistance, conductance, the photoelectric effect, magnetic attraction and repulsion, what a magnetic field is and why electricity and magnetism propagate at right angles to each other, Casimir effect, frame dragging, inverse relationship for gravitational attraction, why objects on the edge of galaxies move at or nearly the same speed as objects closer to the center, diffraction, refraction, reflection, slit experiment, why light propagates rectilinearly and bi-directionally, why the speed of light is constant, and virtually every other phenomenon ever observed or imagined by man. We find that the rope architecture lends itself perfectly to all forms of Push AND Pull, which no discrete particle can ever do.

All these phenomena are made visualizable and therefore understandable without ANY mathematics. Why? Because the language of physics is illustration, not math. Science explains. Description is NOT explanation.

So, the Quantum mechanic has a choice, they must illustrate their objects that they claim are responsible for push and pull, or, they must accept that, just possibly, the physical mediator of phenomena is a continuous structure.

Chapter Twenty Four - Rope Hypothesis

It's one thing to critique theoretical mathematicians and their reifications of abstract verbs into the nouns of reality. It is yet another to offer up alternative, visualizable, understandable, underlying physical mechanisms to perform Mother Nature's tricks. The following chapters are from my books on Bill Gaede's Rope Hypothesis providing a theory of threads for electricity, magnetism, light and gravity. The chapter on the photoelectric effect is from the upcoming book Rope Hypothesis and Thread Theory Part Two.

We've mentioned, previously, the Rope Hypothesis. You can find many chapters devoted specifically to that in my book, Rope Hypothesis and Thread Theory, and also in an upcoming book Rope Hypothesis and Thread Theory Part Two. The remaining chapters in this book are from those books.

Bill Gaede is the originator of the Rope Hypothesis (RH), and writes about it in his book WGDE, available from Lulu and at his website wgde.com. Like me, he is highly critical of what is being called science and especially quantum and theoretical, or mathematical science. It is one thing to be critical of theories, it is another to offer an alternative. Mr. Gaede has done just that. The following chapters are excerpts from my book and the upcoming book on Rope Hypothesis and Thread Theory.

Here it is in brief: The Rope Hypothesis

A single continuous thread weaves in and out of all objects in the universe forming a rope like structure with "knots" or atoms. There are no discrete particles anywhere.

The ropes are comprised of an Electric and a Magnetic thread.

The two (E & M) threads are anti-parallel strands that twist around each other like a DNA strand and separate at atoms.

The M thread wraps around the outside of the atom and the E thread goes to the center of the atom. Atoms expand and contract and the threads unravel, or unwind, reel in and reel out. The pumping action of the atoms is "felt" as a torsion signal along the rope. This is light. The frequency is the number of lengths per unit length. The length of the links changes in atoms of different medium. Amplitude is related to link height.

Rope Hypothesis provides the physical mechanism by which Thread Theory explains gravity and Immediate Action at a Distance (IAAAD).

The tension between objects is the net result of all ropes. This is the force we call gravity (pull).

"As one object approaches another, the EM ropes fan out as a function of decreasing distance and cause the acceleration of one to the other." - BG

Thread Theory explains light: It explains reflection, refraction, diffraction, wave/particle "duality,} wavicles, gravitational lensing, polarization, slit experiments and Olber's Paradox; all atoms are interconnected by a physical medium which mediates light.

TT explains how it is that electricity and magnetism run perpendicular to each other; Anti-parallel electric and magnetic threads wrap helically around each other mediating torsion and tension.

TT explains electric current; Electron serpentines spinning together create what is called electric current. No holes moving left or beads moving right.

TT explains how magnets attract and repel; Sweeping magnetic threads either attract or repel each other similar to jumping ropes interacting.

TT explains why galaxies rotate at about the same rate on the edge as at the center; all atoms are connected by EM ropes.

There is no mysterious Dark Matter or energy.

TT explains covalent bonding; M threads from adjacent atoms spin in the same direction drawing atoms in. E shells merge, there are no discrete electron beads exchanged between atoms.

TT explains ionization; Expanded E shells; no discrete electron particles or electron clouds or orbitals are involved.

TT explains Beta decay; Atoms "pick up" and "drop" crisscrossing ropes called neutrons.

TT explains Ray Reversibility; Interconnecting ropes mediate light in both directions simultaneously. Explains Einstein's light/train Gedanken, and how retroreflectors on the moon "bounce back" the light as the earth and the moon are moving.

TT explains Newton's Laws of motion; "It is inconceivable that inanimate Matter should, without the Mediation of something else, which is not material, operate upon, and affect other matter without mutual Contact…That Gravity should be innate, inherent and essential to Matter, so that one body may act upon another at a distance thro' a Vacuum, without the Mediation of any thing else, by and through which their Action and Force may be conveyed from one to another, is to me so great an Absurdity that I believe no Man who has in philosophical Matters a competent Faculty of thinking can ever fall into it." – Newton

It's all about interconnecting ropes!

TT explains Maxwell's equation; $f = c\lambda$

It is the equation of a rope!

TT explains Mach's Principle and what he meant by; "[The] investigator must feel the need of… knowledge of the immediate connections, say, of the masses of the universe. There will hover before him as an ideal insight into the principles of the whole matter, from which accelerated and inertial motions will result in the same way." – Mach

Those connections are the interconnecting ropes of the Rope Hypothesis.

TT explains why both gravity and electrostatic "force" are stronger or weaker based on "the inverse of the distance squared."

$Fg = G.M1.M2/D^2$

$Fe = K.Q1.Q2/D^2$

Thread Theory explains much more!

Is there anything that TT does not explain? Not that we can find.

Chapter Twenty Five - Forces of Nature
Push and Pull

There are only two forces in nature. Push and Pull. We recognize these forces in various ways and by various names; electricity and magnetism, gravity, strong nuclear force, and weak nuclear force. The Holy Grail of physics has been to unify these forces into one Grand Unified Theory of everything (GUT). While hundreds of GUTs have been proposed, there is only one hypothesis (Rope Hypothesis) that describes the underlying physical mechanisms and its theory (Thread Theory) explains the "forces."

Rational science dictates that we explain all phenomena with objects. Without objects there can be no phenomena. We understand that all phenomena are the result of surface to surface contact between objects. We illustrate, or describe the objects, and we can then explain the phenomena.

We use our senses to observe the world around us, and our intellect to conceptualize the objects and understand phenomena. Our senses are subject to the "forces" which utilize the same underlying mechanisms. Therefore, we have one sense; the sense of touch. This is surface to surface contact between objects. For instance, when molecules touch the olfactory sensors in our nose we have surface to surface contact between these objects, and the result is smell. When we hear, it is a result of surface to surface contact, or touch, between molecules of air and the organ of hearing – our ear. One may read about this in Rational Science Vol. I, Chapter 41.

Although our sensory system is limited to narrow bandwidths, representing only a small portion of everything around us, our intellect (the ability to conceive of concepts) is unlimited. The various objects of the brain which mediate the phenomena we call mind, or thought, awareness, consciousness, and so forth, are neurons, ion channels, synapses, axons, dendrites, etc.

Because there are an unlimited number of combinations and interactions available to the mind, intelligence is not found in the approximately 20 billion neurons and 10^{15} synapses of the brain or in 100 billion neurons of the nervous system, but in the interactions between individual neurons and neural networks. See Rational Science Vol. IV Chapter Seventeen – Tether Hypothesis.

Unless one has a physical problem with their brain, they have this unlimited ability to conceive of concepts, and if they are also smart, they have the ability to apply those concepts towards understanding our reality. The invisible mediators of all phenomena may be conceptualized and understood given that the Rational Scientific Method is applied.

So let's now take a closer look at the "Forces of Nature" Push and Pull. We learned in Rope Hypothesis that spinning electron shells and sweeping magnetic threads account for electromagnetism. We also learned that EM ropes create tension between atoms, and that this tension is known as gravity. Torsion signals created by pumping atoms traveling along the EM ropes between atoms are responsible for what we call light (the entire EM spectrum).

The rope architecture can account for all phenomena previously unexplained. We discussed the irrational paradox of light as a particle and a wave (Rational. Sci. Vol. I, Chp. 11), we discussed Ray Reversibility, Olber's Paradox, and how light travels both rectilinearly and curvilinearly (Rat. Sci. Vol. II, Chps. 15-19).

The rope architecture explains Immediate Action at a Distance (IAAAD), light speed, the rotational velocity of galaxies, and Newton's Laws of motion. We understand that all this phenomena relates to motion, from the motion of vibrating and pumping atoms and molecules that we call temperature, to the rotation of galaxies.

The forces of Push and Pull result in motion at above atomic level of matter. Matter is the set of all objects being comprised of atoms. Let's define motion scientifically:

Motion: two or more locations of an object

Lucretius described Brownian motion perfectly well in his poem, "On the Nature of Things," but the name comes from botanist Robert Brown who discovered pollen grains suspended in water had a strange "jittery" motion. Repeating the experiment using inorganic material, Brown noted that the motion was not due to living objects.

Einstein too was fascinated with Brownian motion and attempted to explain it by way of molecular-kinetic heat action (vibrating atoms). What is temperature but atomic motion? We cover this in Rational. Sci. Vol. I, Chp. 40.

We can imagine motion without pattern as a result of a finite amount of matter not constrained by space; no symmetry, and push/pull from all directions without beginning or end. Pollen diffuses randomly throughout the water due to the irregularity in its shape and size and the resultant irregular gravitational attraction between pollen and water molecules.

Further, the liquid temperature is heterogeneous, meaning that the temperatures vary from location to location. Therefore, some atoms/molecules are vibrating and pumping faster than others.

Motion does not require an observer, nor does motion without pattern require an observer. However, the term pattern seems to invoke an observer, so shouldn't we have a better definition for random? No! We already have a word sufficient for the case, motion.

Brown thought at first that this motion was the result of living objects. Living objects move on their own against gravity. So what is random motion implying? Movement with gravity? It appears that it is not necessary to invoke the term random at all. We simply have motion. That living objects move on their own against gravity is discussed in Rational. Sci. Vol. I, Chapter Forty – Temperature, What Is It?

Monk E. Mind runs through the jungle by moving against gravity. His feet push against the surface of the earth. As he moves through the air, his body pulls air molecules and particulate matter behind him. An ant nearby detects this motion via surface to surface contact between atoms in those air molecules and the mechanico-receptors (hair like structures) on his legs and body. The ant also moves against gravity in response. The earth's gravity pulls against the moon, and both are pulled by the sun.

The sun is pulled by every other object in the galaxy and the galaxy is pulled by the nearest galaxy, Centaurus A and also by Andromeda and every other galaxy.

All this motion is the result of push and pull, which is explained by the connecting ropes, by pumping, spinning and vibrating atoms, sweeping magnetic threads, and endless motion from atoms to galaxies. All objects are continuously pulling on and colliding with other objects.

Quantum Mechanics use push to explain gravity. It is impossible that the ball falls to the floor because it is being pushed away from the floor. Yet, certain "rules" of physics say that within the framework of quantum field theory, graviton and gluon balls "pull" things together somehow (the claim is by imparting reverse momentum) even though they travel away from the source. Gravity according to relativity is no less ridiculous, and is covered in Rational. Sci. Vol. II, Chapter Thirty Eight – Gravity ...Well?

The mainstream idea of "strong nuclear force" is "pull," and "weak nuclear force" is "push," yet there are inconsistencies and contradictions with the current models. A rope-like mechanism can provide both Push and Pull without any contradictions.

Chapter Twenty Six - Gravitation and Electrostatics

Gravitational attraction between two electrons is 8.22×10^{-37} of the electrostatic repulsion at the same distance of separation.

Here's what they use to calculate the ratio between the two:

$k = 8.99 \times 10^9$ Nm2/C2, $e = 1.60 \times 10$-19 C
$G = 6.67 \times 10$-11 Nm2/kg2, $me = 9.11 \times 10$-31 kg

Both gravity and electrostatic *"force"* are stronger or weaker based on *"the inverse of the distance squared."* Even the two have similar formulas:

$Fg = G.M1.M2/D^2$
$Fe = K. Q1.Q2/D^2$

Similarities

"From looking at the two force equations, you can see the similarities and how gravitational force can be considered parallel to the force between two charges.

"Besides being proportional to the inverse of the square of the separation, both forces extend to infinity. They also both travel at the speed of light.

Differences

"One major difference is in the strength of the forces. However, gravitation usually is concerned with large masses, while any large collection of charges will quickly neutralize.

"Another difference between the two forces is the fact that gravitation only attracts, while electrical forces attract when the electrical charges are opposite and repel if the charges are similar. Thus, gravitation is considered a monopole force, while electrostatics is a dipole force." http://www.school-for-champions.com/science/gravitation_electrostatic.htm

Similarities are also used as an exercise in physics courses: http://arxiv.org/abs/1211.2913

If Maxwell's equation is "the equation of a rope for light," then perhaps these equations are the equations of a rope for gravitation and electrostatics (See Maxwell's equation for light).

A single underlying mechanism for both phenomena makes it clear why the similarities and differences, are lost on folks like Franklin Hu who believe gravity is "a straight forward application of the well known electrostatic force."

AND here we find this hilarious example of experimental logic from circlon-theory dot com: "If we simply accept the most basic interpretation of this experiment, then we can easily determine what really happened in that apple orchard over three hundred years ago. After breaking loose from its stem, the apple remained motionless (except for a slight upward acceleration caused by air resistance) while the earth, the air, the tree, and Isaac Newton accelerated upward at the rate of 9.83m/s2, until Newton's head struck the nearly motionless apple as it floated upward on the rising column of air."

Instead of writing 2 equations, simply draw a two atom universe based on the rope model and discover the inverse relationship for electrostatic "force." Draw a 2 object universe with multiple atoms at various distances and see the inverse relationship for gravitation.

What does this describe in regards to the rope model? In other words, what about the rope architecture insures that electrostatic repulsion is greater than gravitational attraction at this micro level, but gravitational attraction exceeds electrostatic repulsion at macro levels of monkEs and men, stars and Galaxies?

The simple answer is that at the macro level there is little or no matter between any two objects for enmeshed E-shells to produce current, or, individual M-threads to interact, but there are still ropes connecting all atoms. At the micro world of, say two adjacent H-atoms, there is only the static tension between the two by virtue of the connecting rope, but E-shells can easily enmesh and magnetic threads can interact.

There are many other interrelations at the level of the fundamental unit of matter.

The theoretical physicist reifies phenomena into particles. The Rope Hypothesis provides an underlying physical mechanism by which Thread Theory can explain all phenomena. Rope Hypothesis provides a model by which we may unite not only gravity and electrostatics, but magnetism and light.

Let's take a look at how these various phenomena are related.

Particle physicists tell us that most of the mass of an atom is attributed to the nucleus. With the rope model, the E-thread extends from each H-atom to the center of all other H-atoms. There will be as many M-threads as there are E-threads contributing to the E-shell. Since electrons are actually NOT discrete, but layers of the E-shell, and M- threads sweep out from the atom, the diameter of the atom is about 145,000 times larger than the nucleus, but the particle theorist attributes more mass to the nucleus.

Listen to this from Wikipedia: "The nucleus is the center of an atom. It is made up of nucleons (protons and neutrons) and is surrounded by the electron cloud."

No wonder they attribute most of the mass to the nucleus, they think the E-shell is a cloud!

If thread is reeled out in one place, slack in thread must be taken up somewhere else. There is a finite amount of thread after all. This can explain inertia and acceleration. What if the slack is being taken up in the sweeping M-threads? In that case, the E-shells would constitute a greater amount of thread than the proton. There is an interrelation of the atoms in any given object. In other words, there is a ratio of H-atoms to other atoms. Perhaps this ratio is maintained by varying pump, spin layers, or by the ratio of pump, spin, and slack between various atoms. Layer(s): thread or threads donated from rope or ropes.

Usually, when we refer to mass we mean the number of atoms comprising all matter, (matter being the aggregate of all atoms).

Mass for an object is the amount of matter comprising it. Mass in terms of Newton's equation is a result of the pull of every atom in the universe via the interconnecting ropes. Here we are engaging in mereology, as well as using our term "mass" ambiguously. This is a result of trying to explain an object, especially the fundamental object which is comprised of thread, rope, atoms, and neutrons as though they were separate entities. Sub-atomic discussions necessarily result in this if there are no discrete particles. We don't explain objects – we point to them and name them, we explain phenomena. However, we are trying to understand what the Particle Phiz Whiz is describing when they invoke sub-atomic particles. They describe phenomena and explain objects. Let's be careful not to do the same. We can use the rope model to explain all phenomena above atom.

If by delving into the inter relations of parts of the fundamental (thread, rope, and atom) we have a better understanding of the Particle Phiz's confusion we have gained something. At least we can see that relationships, that is, ratios and proportions, serve to make us more confident that form and function are often related!

It is clear that the electrostatic and gravitational forces are not confined to objects but also between the electrons and nucleus of a single atom.

The electric force is related by Coulomb's Law as discussed in previous chapters of this book. We leaned that an ion is an atom or molecule with a net electric charge due to the loss or gain of one or more electrons.

AND we understand that charge is not only related to numbers of particles, but is inherent in sub-atomic particles as discussed in the chapter on elementary charge.

With the rope model, charge and elementary charge can be related to direction of spin of atoms, and direction of spin of layers of atoms.

Particle physics is in a quandary: The particle phiz whiz measures the magnetic moment (magnetic field) of an electron and "knowing" its "charge" figures the rate of spin. But this poses a problem. In order for the electron to have the magnetic moment that it does it must be spinning faster (actually called tangential velocity) than light. Since faster than light (ftl) is not possible they conclude that the electron does not spin this fast. In fact, they say, angular momentum alone accounts for the magnetic moment. So, unlike other spinning bodies, sub atomic particles don't actually spin to produce these magnetic fields.

One may say that the rope model has its own quandary:

If the atom is pumping, it is also reeling in and out ropes, whose links must necessarily be spinning. If the atom pumps at c, could not the spinning links be faster than c?

If the center of the atom is spinning at c, how fast is the surface of the atom spinning?

E-shells are comprised of threads coming from every other atom and so create layers in the atom's E-shell that may spin at different rates. If atoms pump and vibrate at different rates depending on their configuration, that is, what object they comprise, could not layers also pump and spin at different rates?

When particle physicists say, "angular momentum alone accounts for the magnetic moment" they are talking about its moment of inertia and its angular velocity. This could actually be one layer of an E-shell, or an E-shell, and would solve their quandary as well as the hypothetical one of the rope model. Do atoms spin at light speed? Various layers of an atom's E-shell might while other layers do not. Perhaps inertia and acceleration is accomplished via the multi-layered E-shells.

We assume that voltage is a result of the number of enmeshed E-shells, and current is a result of rate of spin of these E-shells.

Magnetic strength is a direct result of M-thread density due to alignment of molecules. M-thread density is indirectly related to the rate of spin, as current is related to the rate of spin. How fast an E-shell, or layers of an E-shell are spinning may determine how far out the M-threads sweep.

So we see that magnetic strength is related to the alignment of molecules in a magnet, but an electromagnet has different properties because of the shape of the sweeping M-threads depending on whether "dipole or monopole" structure of the magnet. This is related to shape and nature of the "magnetic field," which is determined by the shape of the magnet (natural or coil or straight wire).

We consider that G (tension) averages out universally by virtue of a constant number of H-atoms, and c (torsion) is also constant because of the inverse relationship between frequency and wavelength which also depends on the constant number of H-atoms. G is a universal average, but gravity locally is a result of the effective pull of ropes.

Effective pull inherently involves inertia and acceleration, and is accomplished via amplitude. Whereas tension is static, amplitude is dynamic. Greater amplitude here must be balanced by lesser amplitude there. E-shells, or layers, of E-shells pumping, or spinning at different rates could be the mechanism of action for this as well.

Coulomb's Law states: "The magnitude of the electric force that a particle exerts on another particle is directly proportional to the product of their charges and inversely proportional to the square of the distance between them. The direction of the force is along the line joining the particles."

As we have proposed, current flowing past a certain point is really a count of how often an E-shell is spinning on its axis at a given point and in a given amount of time. This may be an average of layered E-shells involved.

An individual "particle" would then be a single E-shell or layer of an E-shell. The magnitude of electric force, one layer on the other layer, is directly proportional to their product. The "square of the distance" holds because the E-shell is comprised of a thread, or threads donated from a rope, or ropes whose relationship between frequency and wavelength is inversely proportional. "The direction of force along the line of particles" is by virtue of the E-shells being comprised of a single thread.

The architecture of the rope model insures that no matter what you are talking about something is going to relate either directly or indirectly, and/or either proportionally or inversely proportional to something else! Hence we note gravity and light, electricity and magnetism have the aforementioned relationships.

What I am saying here is that it MAY be that pumping, and spinning are not confined to individual atoms, but also individual layers of the E-shell may pump and spin. Some may be spinning clockwise and pumping at a different rate than others.

Each pumping layer could constitute an electron, but electrons may be whole E-shells gathered from bonding to other H-atoms.

Chapter Twenty Seven - Closed Circuit

David Robison: "Why does a circuit have to be closed to work? If electricity is the spinning of E shells in situ then shouldn't they spin regardless of if a circuit is closed?

"I have a circuit with a light bulb. The bulb is on. I then cut the wire on one side. Why does the bulb turn off? What stops the E shells spinning?"

They only begin spinning in situ (current "flow") when the battery is connected to the bulb.

OK, let's start with the conventional answer: "An electric current is a flow of electric charge. In electric circuits this charge is often carried by moving electrons in a wire. It can also be carried by ions in an electrolyte, or by both ions and electrons such as in a plasma."

Let's replace the terminology with Rope terminology:

Electric currents are enmeshed E-shells spinning in situ. In electric circuits enmeshed E-shells form throughout the power source, conductor and back though a device such as the bulb to the power source.

In this case, the bulb lights if the spinning atoms continue through the whole circuit.

Sometimes you can see the bulb dim and go out, as the atoms stop spinning in the filament. In compact fluorescents you can see the bulb stay lit for a long time after you turn the light off. Why? Because it takes a while for the gas atoms to stop spinning.

David: "So the atoms of the cathode want to spin opposite of the atoms in the anode?"

I don't think so. Most of the atoms need to be spinning together like a drill bit.

David: "When a potential is applied the first E-shell of the circuit spins which induces the next one and so on along the circuit. If the bulb is connected to this wire but the wire is clipped after the bulb leaving the circuit open, what prevents the bulb from turning on still?"

Potential is what electrical engineers use, but how can a potential "do" anything? Also, I think that because there is a continuous rope to atom connection throughout the circuit, that what is happening is happening from both ends simultaneously.

Positivity and negativity depend on which direction you're looking from. Think of a closed circuit like this: (See MonkEs YouTube video on the three atom universe).

On one side it appears as though the rope is spinning one direction. From the opposite side it looks like it's spinning the other direction (A side meaning top of loop versus bottom of loop).

Because of this: (from Rational Science VOl. III, Chapter 39 "Electricity and Magnetism")

No one could agree on which direction electrons flow. Conventional theory had them going from positive to negative and newer electron flow theory (in 1992) had electrons flowing from negative to positive. It is still being discussed today: "When you analyze circuits (i.e. applying Kirchhoff's Laws, etc.), current flows from positive to negative."

"Electrons are negatively charged particles and the flow of electrons (not to be confused with current) is from negative to positive, due to the attraction of the negative charged electrons to the positive terminal. However, current flow is always opposite to the direction of electron flow, and hence it flows from positive to negative in a circuit outside the cell / battery. Within the battery, however, current flows from negative to positive; maybe that was what confused you (the terms within and outside)"

"It really makes no difference, for engineers it is easier to assume it is from + to -."

Wait! What is this?

Electrons travel from the weak battery's positive terminal to the good battery's positive terminal because the weak battery's positive terminal is less positive than the good battery's positive terminal. And since being less positive is the same as being more negative you should be able to see that the current is still flowing from negative to positive. If both batteries had the exact same state of charge then no current and no spark would be produced when you connected them. That's because both negatives would be equally negative and both positives would be equally positive. Just because both terminals are marked + does not mean that one cannot be less positive or more negative than the other one.

http://www.physicsforums.com/showthread.php?t=317139

A battery can charge positive to positive and negative to negative terminals when one is less positive or more negative than the other. Also, as one learns real fast, if not properly grounded, shock from circuit ground can kill you. It is not because of the negative or positive charge, or because of direction of current flow, it is because there is a difference between circuit ground and earth ground. It has to do with the potential difference between two points. We have to have a potential difference and a complete path for current to flow. What is potential? E-shells "waiting" to enmesh.

"It [electric charge] can also be carried by ions in an electrolyte, or by both ions and electrons such as in a plasma."

Of course nothing is being carried anywhere.

When enmeshed E-shells are spinning in the same direction we have "current flow" when adjacent E-shells are spinning in opposite directions and they pull together we may have ionic bonding. Two different ideas about ionization. One we are referring to the "current flowing" through the terminals in the battery and the other the chemical reaction taking place in the electrolytic solution.

Ionization: The M thread never stops spinning around the atom's nucleus, but under certain circumstances it may be induced to swing out further.

It isn't lost to the atom, but it may interact with an M thread of another atom (ionic bonding). So an M thread of a sodium atom interacts with 7 M threads of a chlorine atom and we have ionic bonding...resulting in salt. See Fig. 1 in Bill Gaede's paper http://vixra.org/pdf/1205.0015v1.pdf

Of course, it is understood from Rope Hypothesis that there is neither an electric field nor electrons flowing anywhere! However, a twined EM rope explains the so-called electric and magnetic "fields," and enmeshed E-shells explain current "flow."

Chapter Twenty Eight
Photoelectric Effect and Thread Theory

In this chapter we will examine, however briefly, the photoelectric effect as it relates to the Rope Hypothesis.

The explanatory power of the Rope is unsurpassed in its ability to model reality. There is not a single phenomenon which can not be explained using the rope architecture. Still, this is not why we reject particle physics... not because we think that the alternative is better. It is because the particle/wave paradox is impossible.

We don't reject the mainstream's "explanation" for the photoelectric effect because we think there is an alternative. We reject the "explanation" for the photoelectric effect by particle physics because it is NOT an explanation at all. Please review the Rational Scientific method.

One may claim that the photoelectric effect proves or confirms photons all they want, but photons can not possibly exist! This is covered in great deal in the Rational Science series of books found on Amazon:

"The photoelectric effect is the observation that many metals emit electrons when light shines upon them. Electrons emitted in this manner may be called *photoelectrons*.

"According to classical electromagnetic theory this effect can be attributed to the transfer of energy from the light to an electron in the metal. From this perspective, an alteration in either the amplitude or wavelength of light would induce changes in the rate of emission of electrons from the metal. Furthermore, according to this theory, a sufficiently dim light would be expected to show a lag time between the initial shining of its light and the subsequent emission of an electron. However, the experimental results did not correlate with either of the two predictions made by this theory.

"Instead, as it turns out, electrons are only dislodged by the photoelectric effect if light reaches or exceeds a threshold frequency below which no electrons can be emitted from the metal regardless of the amplitude and temporal length of exposure of light." http://en.wikipedia.org/wiki/Photoelectric_effect

One would reasonably think that if theorists are going to rely on experimental results to confirm their theories then they would have thrown out the theory when their experiments did not confirm it! Instead, they just added particle theory to wave theory and light became a wavicle. It behaves like a wave sometimes and behaves like a particle other times. In fact, they say, it leaves as a particle, travels as a wave and arrives as a particle.

The experiments all claim that electrons are ejected from metal surfaces when bombarded by light. But when one takes a look at the experiments, they only show an effect that is claimed to be caused by the ejection of electrons. How did they arrive at the conclusion that electrons are being ejected? Math!

What the experiments revealed:

1. Energy of electrons increases with light frequency.
2. Current remains constant as light frequency increases.
3. Current increases with light amplitude.
4. Energy of photoelectrons remains constant as light amplitude increases.

So, they measure current, calculate energy and then, based on the math, they extrapolate that electrons are being ejected.

The experiments haven't really changed much as you can determine from this UCLA physics experiment: http://demoweb.physics.ucla.edu/content/experiment-6-photoelectric-effect

They use a photodiode and an amplifier, batteries, a voltmeter, a light source and a filter to vary the light intensity. Shining a light on a metal surface and measuring the voltages at varying frequencies and intensities verifies the theory of particles is responsible; same results, just a new theory. There are basically two components here: voltage level measured, and various results due to light frequency and intensity.

There are other simpler demonstrations you can view on YouTube showing the "photoelectric effect" using static charge to charge tinsel and then ultraviolet light to discharge the tinsel. https://www.youtube.com/watch?v=muxRZ1irsrk

Are these experiments demonstrating that electrons are being "ejected?" Only, if there are light particles called photons and particles called electrons, charge, etc. What does this really show? It shows a simple case of confirmation bias.

As for amplitude, frequency thresholds and work function, that is another matter, and only secondary to the issue of electrons being ejected from metal surfaces. The mechanism is related to the amount of energy (hence frequency) required per given element which increases as the atomic number increases. Frequency, amplitude and Planck's constant are covered with detail in Rational Science Vol. V.

As we have already learned from the Rational Scientific Method, experiments are extra-scientific, and anyways, without a viable hypothesis we have no theory. Since waves and particles fail at the hypothesis stage we have no theory to begin with!

Particles have no way of explaining why higher frequencies result in higher energy. And the higher amplitude light beam does not represent more photons as is suggested, because there are no discrete photon particles.

Energy is the magical word being used here. We are told that higher frequency somehow imparts higher energy to kick out the electrons faster and faster. However, we learned early on in grade school that energy is the ability to do work. Now, these deluded souls wish for us to believe that "ability" is transferred from a zero dimensional particle (photon) to discrete electron balls in the metal plate.

Zero dimensional (photon) particles can not possibly exist and "ability" can not be transferred, so the particle physicist needs to go back to the drawing board and he also needs to abandon his extra-scientific experiments.

But let's briefly look at what is possibly happening using the rope model of light; the same model that explains electricity, magnetism and gravity.

Since all atoms in the light source are connected to all the atoms in the air and in the metal plate, when the light source is turned on it induces the atoms to pump at a rate which results in friction (charge) along the threads of the electron shell. The ropes twist (torsion signal) and also reel in and out between atoms. This is what is known as light.

The number of links per unit length is what is referred to as frequency. The height of individual links is amplitude.

The ropes between atoms whose electron shells are being induced to pump faster have a greater number of links per unit length. Higher amplitude is a result of the superimposition of ropes converging on atoms.

Increased Amplitude: Same number of links with higher height to each individual link. Increased Frequency: faster pumping atom but same amplitude with greater number of links per unit length.

Light shining on the polished metal surface induces electron serpentines to spin. Electron serpentines are enmeshed E-shells of adjacent atoms which turn together like a drill bit, and this electrical current is measured or "felt" in other conductors.

So there you have it, the photoelectric effect without any irrational and impossible, mass-less zero-dimensional photon particles.

Chapter Twenty Nine
Antenna Theory and Rope Hypothesis

Not long ago, during a discussion on antenna theory, I was told this:

"'Fields' near an antenna are physically different from 'far fields'."

The Wickedpedia tells us, "The near field and far field are regions of the electromagnetic field around an object, such as a transmitting antenna, or the result of radiation scattering off an object."

Fields are a where, not a what. So, what we are being told is that a near field is a region of a region near an object. Rational science tells us that we need objects with location in order for there to be any phenomenon. Location alone gets us nothing.

Antenna theory says that the shape of the field is different for near fields as opposed to far fields. Whatever could they be talking about? If a field is a location (a where, not a what), how can it have shape?

There is also talk of fields behaving this way and that, or energy being stored temporarily in magnetic fields, and electric and magnetic fields acting independently of each other, or, one dominating over the other. What does any of this mean?

We can only understand a particular phenomenon by knowing the objects involved. In order to understand electric and magnetic fields, we must be able to visualize the underlying physical mechanisms involved. This is simple with the rope model. Electricity and magnetism are interrelated and may not behave independently of each other because of the rope architecture.

An electric field is a location or region where there are enmeshed electron shells which spin in unison like a serpentine, or a drill bit. This is called current. A magnetic field is where magnetic threads are sweeping around atoms. What is happening with sweeping magnetic threads and spinning E-shells in those locations is what one should be interested in.

Let's just look briefly at a single aspect of antenna theory known as reciprocity.

The "law of reciprocity" tells us that transmit and receive frequencies can be the same (will work equally well). This has to do with current and voltage densities near and between antennas.

In Rope Hypothesis, current strength has to do with the speed of the spinning serpentines, whereas voltage is the number of enmeshed E-shells.

The first Russian Sputnik was always photographed with its antenna facing away (on the other side) so as not to be seen. Had American communication technicians seen the antenna, they could have determined the broadcast frequency from its dimensions.

Know the antenna length and know the number of links of the rope (frequency), as frequency is the number of links per unit length. Find the physical mechanisms involved and one can understand any phenomenon.

Advanced (more than the average persons needs or wants to think about):

The main idea being presented is that only objects have shape, and that the underlying composite "shapes of these fields" are formed from EM ropes. There are lots of things to consider like what accounts for how different antenna types (shapes) propagate more powerfully in specific directions and at specific frequencies? Different antennas propagate more powerfully in specific directions and at specific frequencies.

Though the radiative fields die off at $1/r$, the power density falls off at $1/r^2$. There are also reactive fields such as the H-field and E-fields. The rope model can offer explanations that were never addressed in "theory" class.

I forget most of what they taught me in my training as a Line Of Sight (LOS Microwave) radio mechanic in the military. It has been decades, and anyways the theory just pertained mostly to math and other abstractions that I wasn't interested in committing to memory beyond taking the daily and weekly block tests. The practice "in the field" was more meaningful than discussing E and H fields. Frying a buzzard on a high line, or exploding a can of Coke was far more rewarding than computing field strengths, etc.

A brief perusal of antenna theory reminds me of this:

The radiating field falls off at $1/r$. This likely has to do with the superposition of EM ropes.

An E (electric) field is measured at a specific point in volts per meter and falls off at $1/r^2$ whereas an H (magnetic) field is a vector quantity (both magnitude and direction), is measured in amps/meter and falls off at $1/r^3$.

Measuring "coulombs" at a specific point is predicting how fast one E-shell is spinning in place, NOT telling us that a single electron is flowing past a particular point. The magnetic field vector is still about measuring a spinning E-shell but extrapolating along multiple paths (angles) or many ropes around a specific location. As E-shells spin the magnetic threads sweep out around the atoms.

The propagating waveform is the interaction between the two. With the propagating waveform the E and H fields are perpendicular but in phase with each other. In other words, the E and M threads are wrapped helically around a common axis.

Near the antenna (near field) the H and E fields are 90 degrees out of phase. E fields are plotted at points along a straight path. M fields are considered along multiple contrived curved paths out of every possible path in all directions, so this averages out to 90 degrees out of phase.

Between the near and far field is a region called the radiating near field where the reactive fields no longer dominate and the shape varies greatly with the distance. These regions probably equate to the "Bird's Beak," linear, and exponential regimes of Thread Theory. In other words, rope density.

A highly directive antenna will be large, or many wavelengths in height, and a low directive antenna will be small. This has to do with current and voltage distributions. Why this is, has never been explained, just described mathematically. Of course, when viewed through the lens of Rope Hypothesis we can see that a taller antenna will provide many more rope paths (more angles = more wavelengths) to and from the "target."

The far field ratio of electric and magnetic fields is a direct result of the ratio of these fields in the transmitter, but absorption in the far field does not feed back to the transmitter as it does in the near field. Why is this?

Feedback is nearly instantaneous and continuous as the ropes are a two way street as far as any and all signals are concerned. Absorption in the far field region does not put a load on the transmitter because it does not feed back with any strength (amplitude) due to the superposition of the ropes in the near field radiating region. Absorption simply means that signals have been shifted out of the tuning range of the transmitter and receiver.

Near field absorption puts a load on the transmitter, however, as many separate paths are affected at once and since the signal strength has not yet fallen off, the feedback is felt at the transmitter with greater consequences.

An important Note: Applying the rope model to Antenna theory is the Thread Theory of Antennas.

Don't forget to apply the Rational Scientific method to everything.

What should one conclude from this chapter?

That "wave propagation" and "fields" are merely descriptions of the electric and magnetic threads forming ropes between atoms in the transmitter and receiver and all points in between. Without the Rope Hypothesis we have no explanation, or Theory of Antennas.

Chapter Thirty
Light, Gravity and Magnetic Moment

My friend David Robison said, "Gravity falls off exponentially with increased distance. Light falls of exponentially with increased distance. Of course! Both light and gravity operate on EFFECTIVE ropes." This sparked an interesting discussion related to the Rope Hypothesis and how it is used in a theory of threads explaining the various phenomena.

Discussions of magnetic fields often refer to them as though they behave like light. "Does magnetic field strength follow an inverse square law?"

According to the Naked Scientist website, "Yes--the decay of $1/r^2$ comes from it basically being light-like electromagnetic radiation."

They are talking about "magnetic field strength around current carrying wires." A natural magnet's strength decays at $1/r^3$ instead of $1/r^2$. A hysteresis coil at a distance produces a magnetic field strength which decays at $1/r^3$, if I recall correctly from my TV repair days.

The difference is what they are calling monopoles and dipoles. One will often hear something like this from WIKI: "the magnetic field due to a dipole is inverse cubic, but the magnetic force from a monopole is inverse fourth order."

Magnetic moments, dipoles, monopoles; what does all this mean? Light and gravity conform to the hairless ape's "inverse square law" but magnetism to an "inverse cube law". Both are results of the Rope architecture, and since electricity and magnetism often are two sides of the same coin, we have to understand the relationship.

Whereas gravity "falls off exponentially with increased distance" because of "effective pull" and light "falls off exponentially with increased distance" because of changes in frequency and amplitude, magnetism falls off as the square or the cube depending on the source (current/monopole or magnet/dipole) and shape of wire/magnet.

Provided there is nothing between the observer/detector and the light source to change the frequency, the ropes still superimpose with distance (as with gravity) and the "effective light" decreases.

Current along a wire (read, enmeshed E-shells spinning in situ) along with the accompanying M-threads result in the field strengths detected. Speed of rotation results in a wider path of swinging threads. More magnetic molecules in a natural magnet, or more enmeshed E-shells (greater current) result in greater M-thread density. So, at near proximity to a source the numbers of threads per area are greater than distally.

The magnetic field strength detection and equations approximate the dipole fields falling off as the cube of the distance. Close in to a magnet, the proximal end can be detected/calculated and the distal end treated as if a monopole. In this case the field falls off as $1/r^4$ (quadractically).

Really close to one end of the magnet the fields effectively behave linearly.

Calculated or detected strength depends on location and shape of the source.

Reading the wiki article on magnetic moment and taking in their illustrations we see that an equation is used to calculate a magnetic moment. Magnetic moment is a vector quantity (magnitude and direction) derived as a ratio of torque to magnetic field strength.

From wikipedia, on magnetic moment: "The sources of magnetic moments in materials can be represented by poles in analogy to electrostatics. Consider a bar magnet which has magnetic poles of equal magnitude but opposite polarity. Each pole is the source of magnetic force which weakens with distance. Since magnetic poles always come in pairs, their forces partially cancel each other because while one pole pulls, the other repels. This cancellation is greatest when the poles are close to each other i.e. when the bar magnet is short. The magnetic force produced by a bar magnet, at a given point in space, therefore depends on two factors: the strength p of its poles (magnetic pole strength), and the vector l separating them. The moment is related to the fictitious poles as U=pl."

They use this as an analogy, but are closer to reality than they realize. The confusion for these guys is that they believe in spinning particles and electron orbits. They do not understand the relationship between E and M threads, although they observe (or calculate) the effects.

The relationship of magnetism to electricity is the serpentine drill bits (enmeshed E shells) along with the accompanying jump rope like M-threads. For discussion about this with the luxury of detail, please go to the Rational Scientific Method or Rope Hypothesis facebook group.

Opposite charges are oppositely spinning E-shells along with sweeping M-threads.

Comparing the bar magnet to the electromagnetic loop in Wiki's image we see the same field strength and shape but at a different orientation.

Without an underlying physical mechanism like one the Rope Hypothesis provides, they are destined to observe, describe, measure and calculate forever without really understanding what's happening.

As long as they continually invent particles and orbitals they will forever be confused and doomed to stating things like this: "The preferred classical explanation of a magnetic moment has changed over time. Before the 1930s, textbooks explained the moment using hypothetical magnetic point charges. Since then, most have defined it in terms of Ampèrian currents. In magnetic materials, the cause of the magnetic moment is the spin and orbital angular momentum states of the electrons, and whether atoms in one region are aligned with atoms in another."

Of course, in order to "explain" these "point charges" and "states of electrons" they have to chant the mathemagical incantations of Quantum Magic.

Yes, they are very good at describing what they detect with their devices since the devices are calibrated and interpreted using their mathematical formulas. As the mountain said in "Me and My Arrow", "You see what you calculate to see and hear what you calculate to hear."

When they say this: "The net magnetic moment of any system is a vector sum of contributions from one or both types of sources. For example, the magnetic moment of an atom of hydrogen-1 (the lightest hydrogen isotope, consisting of a proton and an electron) is a vector sum of the following contributions:

- the intrinsic moment of the electron,
- the orbital motion of the electron around the proton,
- the intrinsic moment of the proton."

What they are really describing is this: The effect of magnetic threads on our detectors is a calculated sum of thread density and direction of magnetic threads sweeping around an axis.

When they say this: "Similarly, the magnetic moment of a bar magnet is the sum of the contributing magnetic moments, which include the intrinsic and orbital magnetic moments of the unpaired electrons of the magnet's material and the nuclear magnetic moments".

Restated they really mean this: The atoms and molecules of a magnet align themselves increasing thread density sweeping in two opposite directions from a division we term north and South Pole.

Instead of this: "Viewing a magnetic dipole as a rotating charged particle brings out the close connection between magnetic moment and angular momentum. Both the magnetic moment and the angular momentum increase with the rate of rotation. The ratio of the two is called the gyromagnetic ratio and is simply the half of the charge-to-mass ratio."

We can say this: A spinning atom's magnetic threads extend out from the atom's E-shell and sweep out around its proton's axis with more threads participating the greater the number of atoms involved. We can propose a ratio of M-thread density to speed of sweeping M-threads, and call it the gyromagnetic ratio.

Conclusion: Whereas gravity and light "strength" are directly related to effective numbers of ropes, magnetic field strength is directly related to effective M-thread density and the speed of their sweeping action.

The shape of the magnetic source determines how many effective threads. Ropes superimpose over distance and M-threads combine or effectively cancel each other out depending on the shape of the source.

Rope Hypothesis provides the model by which Thread Theory can explain any and all phenomena. Read the books Rope Hypothesis and Thread Theory by Monk E. Mind (found on Amazon), and WGDE by Bill Gaede (found on Lulu and available from his website wgde.com. Come by the Rational Scientific Method and The Rope Hypothesis Facebook groups to discuss any and all topics related to the Rational Scientific Method and Rope Hypothesis and Thread Theory.

More books by Monk E. Mind found on Amazon

Check out the Rational Scientist Magazine at :

www.rationalscientist.com

Check out Monk E mind books and reviews:

www.monkemind.com

From Short Shorts, Short and Very Short Stories:

End of The Trail

He walked along the trail with all the other workers. They had toiled all day in the field, and now were heading back to join the rest just over the hill. His kind had lived and worked this land for over a thousand years. They are the hardest workers anyone has ever known.

They were all tired and hungry, and it was quiet as they mindlessly shuffled down the trail. He had walked this way many times before, as they all had, without a single thought about the individual sacrifice each has made for the collective. This is the way it has always been. His large strong body moved forward with no thought about what tomorrow would bring. In fact, he didn't think anything at all. None of them did.

Suddenly a bright white intensely hot beam of light shot out of the sky. His legs curled up underneath him as he collapsed, instantly dead. His insides were cooked and a single puff of smoke rose from his body with a pop.

"Time to eat" Jimmy's mother called from the back porch. Jimmy put his magnifying glass in his pocket and muttered under his breath, "Stupid ants".

Monk E. Mind

From the Novella, Bug World

Bug World

This story is a work of fiction. The characters, incidents and dialog are drawn from the author's imagination and are not to be construed as real. Any resemblance to actual persons, living or dead is purely coincidental. The insects, on the other hand, are real.

The Hardwick family pulled out of the caliche driveway leaving their modest four bedroom ranch style home behind a cloud of white powder. They made their way down Haines County Road 41 and headed towards the Interregional.

Mr. Henry Hardwick was editor of the local newspaper, the San Martin Daily News. His wife, Julie, was principal of San Martin Middle School. The two kids, Jimmy and Sally, were playing happily in the back seat of the Force Trax Cruiser.

They quickly passed through their small Central Texas town situated in the middle of Haines County and turned on to IH 36.

Today was Saturday and they were going to Bug World. Jimmy was finally going to get his ant farm and Sally a butterfly garden. Henry was getting a break from yard work and Julie was sure she was going to be getting the willies.

"Don't worry, dear, the website says that there is a gift shop and a lounge where they serve soft drinks and tea. The scary bugs are down in the basement and the more common bugs up stairs where the entomologist will be starting the tour. You can relax in the lounge with the other squeamish ladies," Mr. Hardwick assured his wife.

"Right, like I'm going to be relaxed with a zillion creepy crawlers all around me," Mrs. Hardwick said half-jokingly. "And what makes you think it will be just ladies? Remember when that cockroach flew right at you and landed on your forehead? You screamed, and danced, and waved your hands in the air like you were being eaten alive!"

Jimmy and Sally laughed at that, then Jimmy teased his mom and dad. "Better be careful… insects can smell fear."

Jimmy chuckled and returned to his SimAnt game on his Game Boy Advance. Sally continued brushing the mane and tail on her My Pony.

Henry disliked ants, mostly because they were such pests. The last year had been especially bad with dozens of ant mounds popping up in the yard almost daily. He was continuously pouring boiling water on mounds, poisoning, and otherwise killing off ants. It took up far too much of his time.

Henry didn't like using pesticides. Julie hated killing anything, including bugs. They were always trying to find a compromise, but these ants were more than a nuisance, it felt like fire when they bit you.

One time Henry cut the top off of a Honey Bear. It had about a third of a bottle of honey left in it. He placed it in the back yard near a busy nest. Ants climbed in and they couldn't climb back out! The next morning the Honey Bear was completely full! Unfortunately, there was an endless stream of ants trailing in from nearby nests.

Jimmy knew quite a bit about ants. He had learned a lot from SimAnt info files. He told his dad that he thought he could solve the problem once and for all if his dad would let him handle it. Henry agreed and told Jimmy, "If you can solve our ant problem, I'll take you to Bug World and you can get the ant farm you've been wanting while we're there."

Amazingly, within a few weeks there were no noticeable ant mounds in the yard and only a few ants here and there foraging. There were a few piles of ant bodies, but no other signs of ants. There were no new mounds popping up either!

"How did you manage this, Jimmy?" his father asked totally in awe of his son's accomplishment. "Well, dad, ant colonies compete for food and water. I killed off all the mounds but one at the waterfall and Koi pond so there is only one large healthy colony of ants. They'll keep the rest out from now on."

Today Jimmy would be getting his Ant Farm with 25 Western Harvester ants and their queen, as promised. It was only fair that his sister got something too. Sally was getting a live butterfly kit with Painted Lady Butterfly Caterpillars. She would release them as adults and let them fly away, of course.

"Speaking of insects, Henry, there have been an unusually high number of absentees lately. The notes from home are saying it's because of various insect bites, spiders, chiggers, ants, fleas, ticks… and we've had a real problem with head lice lately" Julie Hardwick said.

"Hey kids, look, we're almost there!" Henry exclaimed enthusiastically. Everyone started reading the signs in unison in a sing-songy voice:

"Kids Bugging You?"

"Tired Of Driving?"

"Then Stop at Bug World"

"You'll Soon Be Arriving!"

Bug World was an entomologist owned roadside tourist stop fronting the access road to IH 36. It was a Civil War era home that had once belonged to a wealthy rancher and banker. The sign said that it was currently owned by "Dr. Harold 'Bugs' Smithton, Proprietor – Entomologist." The brochure said Smithton specialized in social insects; ants, bees and wasps. He converted the old house into Bug World because he wanted to educate the public on insects, and he needed an office and laboratory to carry on his work. He was often called upon to help the local sheriff and he also worked with the forensic pathologists and coroners in the county when they had difficult cases.

Although his formal training was mostly in myrmecology, he had become known around the county for his ability as a forensic entomologist after helping to solve the murder of poor old Mr. and Mrs. Howard.

During the week he helped at the nearby Body Farm, taught an occasional class as a guest speaker in University Entomology departments, worked with the Agriculture Extension Service, and offered testimony as an expert forensic entomologist in various court cases. On the weekend he opened up Bug World to the public.

The old mansion had lots of Queen Anne features, including an eight sided turret with an onion dome roof. Its arched windows and curved porch with the spindlework balustrade and overlapping shingle work gave it a striking appearance reminiscent of homes that were built 200 years ago.

Bug World was a large two-story house built in the 1880's. It was an L form house with a complex roof and corbelled chimney popular during the time it was built. It had sexagonal turrets in the front with a conical gabled roof over the porch.

The back part of the house was covered by a gabled roof and had kicked, or, extended eaves. It had beautiful round arched windows on the ground floor and keyhole windows on the second story with lots of stained glass everywhere.

Mr. Hardwick pulled into the parking lot and had barely stopped before the kids spilled out onto the pavement. Taking charge, Henry said, "Let's all stretch a moment and go over a few things before we go inside."

Henry laid down the law about staying together, not talking during the tour unless acknowledged by the tour guide, not running and not touching anything. If the kids fully complied,

they would be getting their Ant Farm and Butterfly Garden on the way out afterwards.

Jimmy could barely contain himself and shifted from one leg to the other impatiently until his dad stopped talking. When given the go ahead, Jimmy and Sally darted towards the front door. "Kids! What did I say?" hollered their dad. Jimmy and Sally slowed to a crawl until mom and dad caught up to them.

The Hardwicks entered into Bug World and found themselves on the first floor where the gift shop and ticket counter was located.

The walls were lined with displays of beautiful butterflies, racks of jewelry made from beetles and other insects, collection nets and jars, thousands of bugs sealed in plastic domes, ant farms, butterfly gardens, videos and more.

The sound of Burl Ives was heard over the speaker system singing the Ugly Bug Ball...

"Come on let's crawl (gotta crawl gotta crawl)
To the ugly bug ball (to the ball to the ball)
And a happy time we'll have there, one and all at the ugly bug ball."

A nice young lady named Frances greeted them all as dad shelled out the five dollars each for the admission price. "Welcome to Bug World. Dr. Smithton will be leading a tour starting on the top floor in a few minutes. I'll announce that shortly. Please feel free to peruse the gift shop and do not hesitate to ask me about anything that we have, but save your questions about bugs for your tour guide. There are drinks and snacks in the lounge right through that door. Enjoy your visit."

Antz, Dreamworks/Pixar animation movie, was playing on a large screen TV. The critters were on the march towards the termite colony, and Danny Glover's voice was heard to say, "Overwhelm their defenses, and kill their queen." Woody Allen's character replied "Hey fellas, that's being a little extreme, I feel. Why don't we just try to influence their political process with campaign contributions?"

After lots of ewwws and ahhhs over scorpion necklaces, Jewel beetle rings and bug posters, it was announced that the tour was about to begin. Everyone lined up at the bottom of the stairs except for Julie and a few others who remained in the lounge. Julie parked herself on a settee next to a woman who had brought two of her grandchildren.

Frances removed the rope that was stretched across the bottom of the stairwell between the newels and let it hang down from one of the beautiful volutes. She politely asked everyone to please use the handrails on their way up to the second floor.

A tall thin gentleman with bifocal glasses and a bow tie stood at the top and waited patiently as the group of 15 visitors gathered near the landing.

"Hello, and welcome to Bug World. My name is Harold 'Bugs' Smithton, you may call me Bugs." The children all chuckled, and a few parents whispered, "You will call him Dr. Smithton."

"I am your tour guide today. I will be showing you around the facility and answering all your questions. Before we take a look around, I would like to ask you a question. Does anyone know what entomology means? You, young man," said Mr. Smithton pointing at Jimmy.

Jimmy lowered his hand and proudly announced, "Entomology is the study of insects."

"Very good. That's correct. And I'm an entomologist, whose specialty is in myrmecology and other eusocial insects. A myrmecologist studies ants, and eusocial insects are those insects that are highly social. They cooperate by raising their young, the old and the young live together with multiple generations, and there is a division of labor. For instance, ants have queens, drones, soldiers and workers."

"Instead of using all the big words associated with binomial nomenclature ...er... the two names corresponding to genus and species, I'll use the more common names or refer to a particular order by what it means. For example, grasshoppers, katydids and crickets are from the order Othroptera, which means straight winged. And, of course, it's far easier to remember or say 'grasshopper' than 'Melanoplus differentialis.' You'll find the scientific names on the label below the common names on each display or terrarium."

"Bugs are interesting. It's why you are here, isn't it? Did you know that bugs are also beautiful, strong, resourceful, useful...and necessary?

"Not all insects are bugs! Bugs are a type of insect! True bugs have a mouth sort of like a straw, called a stylet. It's O.K. to call insects bugs. It's a common misconception. Spiders aren't insects, they are arachnids. Scorpions, mites and ticks are also arachnids. Bugs, insects and arachnids are all arthropods, which means they have their skeleton on the outside of their bodies. Scientists classify things in a certain way that helps them to organize and to understand how they are related. "This is really important because there are at least a million insects that we know about and some

scientists think there may be as many as 30 or 40 million more that we don't know about.

"There are many interesting facts about insects and bugs. There are many things that most people do not understand or

appreciate about insects and bugs. That is why Bug World is here.

"We should learn to get along with them, for one reason because there are so many of them. In the country, there are more of them in one square mile than there are human beings on earth.

"If we put all the insects on one end of a scale and all the other earth creatures on the opposite end, insects would outweigh everything else. There are more than 200 million for each human. Termites alone outweigh humans ten to one! "Insects make up nearly 80% of all Earth's animals, and there is a reason for that as we shall see shortly.

"Do you think humans can run fast? The tropical cockroach can run 50 body lengths in a second. This would be like a human running the 100 yard dash in one second. That is nearly 200 miles per hour!

"Do you think humans are strong? Many insects can lift 50 times their own body weight. This would be equivalent to a human lifting a car full of passengers. A Rhinoceros beetle can lift over 800 times its own weight!

"Do you think humans can jump high? A flea can jump over 100 times its own height, which would be like a human jumping over the Washington Monument!

"Do you think humans are great survivalists? A cockroach can live for a month without food and can live for nine days without its head. One species of midge fly can live for three days in liquid nitrogen which is minus 321 degrees Fahrenheit.

"Not only are insects interesting, they are useful to humans too. Bees provide honey, beeswax, pollen, propolis and royal jelly

for human use and consumption. Their venom has been known to be useful in treating arthritis and multiple sclerosis.

"For several thousand years, Carpenter ants have been used like stitches to close wounds in some countries. Some maggots are used as a medical therapy to eat decaying flesh on a wound thereby preventing infection and sometimes saving a limb from amputation.

"Beetles are used for jewelry because of their beauty, silk worms for making fabric and bees are used as pollinators. The Australians regularly import dung beetles to help rid them of cattle poo. The dung beetles significantly reduce methane gas too.

"Lady bugs and Green Laceleafs are excellent for biologic pest control. Lac Scale insects are used for polish, varnish and printing ink. Insects are used for dyes and for food. Did you know that beetles taste like apples, wasps taste like pine nuts, and worms like fried bacon. Worms and other insects are helpful for catching other food like fish.

"Entomologists can use information about the life spans and habits of insect for solving crimes, or determining time of death of livestock or humans. Gill Grisholm of C.S.I. and T.J. on Bones are both forensic entomologists.

"Scientists use fruit flies for genetic research. There are many more uses man has come up with for insects, but nature depends on insects for things like breaking down waste, wood and plant matter, as well as recycling animal bodies.

"Over the next forty minutes Harold 'Bugs' Smithton told the wide eyed children and curious adults many things about insect life, their habitats, and also fielded many questions. They saw fascinating insects both live and preserved. Some of the older children and adults went down into the basement where they saw the scary ones like the giant scorpions, vinegaroons, Bot flies, hornets, carrion beetles, and maggots and ticks at work.

Harold Smithton ended the tour with this exhortation, "Over the last hour we've seen how interesting, useful, and necessary bugs and insects are. So, the next time a bug or insect crosses your path, I hope you'll recognize the difference and that you'll think twice before squashing it!"

The Hardwicks spent another hour in the gift shop, and on the way out they purchased the kid's Ant Farm and Butterfly Garden. Henry bought some green Laceleafs and ladybugs for biologic pest control, and after a long drawn out discussion Jimmy was allowed to get the living dragonfly on a string with his allowance.

"You'll fly that thing on a short leash, young man!" his mother told Jimmy. His father added, "Keep it in its cage until we get home. I don't want us to be another highway statistic." Henry recalled a recent article in the Daily News about highway accidents, informing readers that insects were among the top twenty causes of motor accidents.

The kids were very pleased with their acquisitions and were singing a song from 'It's A Bugs Life.'

"It's a Bugs World After All

...the morning dew falls down on our fuzz...
...underneath a leaf or behind a stone..."

Julia told Henry about her conversation with Mrs. Hansen, the grandmother she met in the lounge. It wasn't really much of a conversation. Julia mostly sat there politely listening to Mrs. Hansen's many stories about her children, her grandchildren and how she rarely got to see them, the late Mr. Hansen, the neighbor's dog that never stopped barking, the bridge club and on and on it went.

"The old girl did have one very interesting story I thought you'd want to hear." Julia continued, "She and the grandkids drove up from San Antone. Mrs. Hansen lives in the country a little south of there and said that folks around those parts were having lots of problems with ants. She has 12 bird houses and the ants were killing off the baby birds. It had gotten so bad that the black-capped vireos which lived there had all flown off to 'who knows where?' Reports of baby rabbits and small deer being eaten alive were almost a daily occurrence and it was creating quite a stir among the country folk."

"Mom, how many miles per hour is 58 kilometers per hour?" Jimmy asked. His mom responded, "About 35, dear. Why do you ask?"

"Because the brochure about Elliot says that he can fly that fast. It also says that dragonflies are not really flies even though they have fly in their names, and the largest dragonfly was a prehistoric one called Mega...new...rah...mon...yi, and they can have as many as 30,000 lenses in each eye, and they have six legs but can't walk, and the helicopter design was inspired by the dragonfly...and..."

A Curious Chain of Events

Published in New Worlds, 1939, Arthur C. Clarke responded to the editors, authors, and fans who bemoaned, "All ideas in science fiction have been used up," with this statement: "Notwithstanding the pessimists, there are a million themes that science fiction has never touched."

He went on to say, "As long as science advances, as long as mathematics discovers incredible worlds where twice two would never dream of equaling four, so new ideas will come tumbling into the mind of anyone who will let his thoughts wander, passport in hand, along the borders of Possibility."

Arthur C. Clarke was close, little did he know that he was predicting how science would turn out, not science fiction. Today we see that the lines are blurred between science fiction and scientific "Possibility."

"2+2=5 for extremely LARGE values of 2." – Lawrence Krauss, Theoretical Physicist, Cosmologist

Today, we live in a world where two plus two can equal five. Theoretical physicists weave fantastic yarns of multiple dimensional realities, mass-less particles with infinite densities, and numbers that can add up to anything.

Science has replaced magic and superstition, with Quantum Mechanics and Superposition, magical incantations with mathemagical formulas, supernatural deities with multi-dimensional realities, and God Creation with Big Bang Creation. It is a great time for science and for science fiction!

A Curious Chain of Events seeks to address the main issues of science and the philosophy of science. What is the difference between an object and a concept? What is truth and love? What does it mean to be alive, to have consciousness and intelligence? What is reality? What does it mean to exist? What is motion and what is time?

A Curious Chain Of Events is a story about the love between a man and his wife. Paul is a scientist and Rhonda was a philosopher. Paul is grieving over the loss of his wife who died recently in an auto accident. Rhonda is always on his mind, and he attempts to hold on to her memory by obsessing over their previous conversations. They were often on opposite sides, but Paul had learned to respect, and even depend on Rhonda's point of view. Paul comes up with many ideas while reliving their discussions, and is working on a number of projects at his company's laboratories centered mostly on synthetic biology, DNA, artificial intelligence, and various single celled organisms. Paul will do anything to keep Rhonda's memory alive. Anything.

Paul's company, LoTech Laboratories, has developed a new type of photo voltaic, oil eating bacteria, pollution absorbing phytoplankton, Large Cell Technology, Smart Jeans (yes, with a J) and more. Kept secret from the rest of LoTech Labs, Paul and a small team have created an artificial intelligence that lives in a neural network in a hidden lab. The story brings to the reader's attention many philosophical ideas and ethical concerns about the sanctity of life and bio technology.

Although the story is fiction, it is based on cutting edge science and technology. Be prepared to stretch your imagination and your reading comprehension. Expect to be challenged intellectually and philosophically, to expand your borders, and to possibly step out of your comfort zone.

Rational Science Vol. I

Chapter One – The Rational Scientific Method
Chapter Two – Existence
Chapter Three – Authority
Chapter Four – Truth
Chapter Five – Mathematics
Chapter Six – Love
Chapter Seven – Space
Chapter Eight – Expanding Universe
Chapter Nine – E=mc Squared Away
Chapter Ten – Einstein
Chapter Eleven – Light, Particle Or Wave?
Chapter Twelve – Smoke and Half Silvered Mirrors
Chapter Thirteen – String Theory
Chapter Fourteen – Time Part One
Chapter Fifteen – Time Part Two
Chapter Sixteen – Time Part Three
Chapter Seventeen – Mass Part One
Chapter Eighteen – Mass Part Two
Chapter Nineteen – Mass Part Three
Chapter Twenty – Quantum Magic
Chapter Twenty One – Big Bang Theory
Chapter Twenty Two – Black Holes
Chapter Twenty Three – What Happened To the Dinosaurs
Chapter Twenty Four – Math Crazy
Chapter Twenty Five – History of the Rational Scientific Method
Chapter Twenty Six – Neo Mechanical Gravitation Theory
Chapter Twenty Seven – Source Theory
Chapter Twenty Eight – The Electric Universe
Chapter Twenty Nine – Infinite Universe
Chapter Thirty – Creation Science Isn't
Chapter Thirty One – Intelligent Design Isn't
Chapter Thirty Two – Emergent Complexity
Chapter Thirty Three – Creation Vs Evolution
Chapter Thirty Four – Darwin's Black Box
Chapter Thirty Five – Dowsing
Chapter Thirty Six – Free Electricity & Dennis Lee

Chapter Thirty Seven – Mighty Engine
Chapter Thirty Eight – Homeopathy
Chapter Thirty Nine – The Holofractographic Universe
Chapter Forty – Temperature, What Is It?
Chapter Forty One – The Sense Of Touch
Chapter Forty Two – Life
Chapter Forty Three – Optical Illusion Glasses
Chapter Forty Four – Peer Review

Rational Science Vol. II

Chapter One – The Rational Scientific Method
Chapter Two – Scientist, Science, & The Scientific Method
Chapter Three – Scientific Method? For Dummies!
Chapter Four – Hypothesis, Theory, Conclusion
Chapter Five –Science & Technology - Conceptual & Empirical
Chapter Six – Experiments Are They Part of the Scientific Method?
Chapter Seven – Pseudo-Scientist Index
Chapter Eight – Proof Is For Alcohol
Chapter Nine – Rational Thinking Test
Chapter Ten – Mind Science
Chapter Eleven – Matter and Motion
Chapter Twelve – Dimensions
Chapter Thirteen – Dimensions of Reality
Chapter Fourteen – The Three Dementia of Geometry
Chapter Fifteen – The Nature of Light
Chapter Sixteen – Light ...Does It Travel Rectilinearly or Curvilinearly?
Chapter Seventeen – Distance To the Stars
Chapter Eighteen – Shapiro Effect
Chapter Nineteen – Distance ...The Rubber Ruler
Chapter Twenty – Atoms
Chapter Twenty One – The Four Quark Circus
Chapter Twenty Two – Does the Atomic Bomb Work?
Chapter Twenty Three – Abiogenesis

Chapter Twenty Four – Adaptive Mutation
Chapter Twenty Five – Computer Simulation Argument
Chapter Twenty Six – E=mc^2
Chapter Twenty Seven – True Story - God vs. Science; Food For Thought
Chapter Twenty Eight – Zero Point Energy Nonsense
Chapter Twenty Nine – Global Warming Lie
Chapter Thirty – The James Webb Space Telescope
Chapter Thirty One – JWST and the BBT
Chapter Thirty Two – Stars, Protoplanets, and Planetesimals
Chapter Thirty Three – James Webb Space Telescope and Extra-Terrestrial Life
Chapter Thirty Four – Gravitational Lensing and the CMB
Chapter Thirty Five – Dark Energy and Dark Matter
Chapter Thirty Six – Interstellar Space Travel
Chapter Thirty Seven – Size, Does It Matter?
Chapter Thirty Eight – Gravity ...Well?
Chapter Thirty Nine – Electricity and Magnetism

Rational Science VOl III

Chapter One – The Rational Scientific Method
Chapter Two – Scientist, Science, & The Scientific Method
Chapter Three – Scientific Method? For Dummies!
Chapter Four – Hypothesis, Theory, Conclusion
Chapter Five – Science & Technology - Conceptual & Empirical
Chapter Six – Experiments Are They Part of the Scientific Method?
Chapter Seven – Proof Is For Alcohol
Chapter Eight – Pseudo-Scientist Index
Chapter Nine – Peer Review
Chapter Ten – Mathematics
Chapter Eleven – Existence
Chapter Twelve – Authority
Chapter – Thirteen - Truth
Chapter Fourteen – Space
Chapter Fifteen – Expanding Universe

Chapter Sixteen – E=mc Squared Away
Chapter Seventeen – Light, Particle Or Wave?
Chapter Eighteen – Smoke and Half Silvered Mirrors
Chapter Nineteen – String Theory
Chapter Twenty – Time Part One
Chapter Twenty One – Time Part Two
Chapter Twenty Two – Time Part Three
Chapter Twenty Three – Mass Part One
Chapter Twenty Four – Mass Part Two
Chapter Twenty Five – Mass Part Three
Chapter Twenty Six – Big Bang Theory
Chapter Twenty Seven – Black Holes
Chapter Twenty Eight – What Happened To the Dinosaurs
Chapter Twenty Nine – Life
Chapter Thirty – Creation Science Isn't
Chapter Thirty One – Intelligent Design Isn't
Chapter Thirty Two – Emergent Complexity
Chapter Thirty Three – Abiogenesis
Chapter Thirty Four – Adaptive Mutation
Chapter Thirty Five – Creation Vs Evolution
Chapter Thirty Six – Darwin's Black Box
Chapter Thirty Seven – Temperature, What Is It?
Chapter Thirty Eight – The Sense Of Touch
Chapter Thirty Nine – Dimensions
Chapter Forty – Dimensions of Reality
Chapter Forty One – The Three Dementia of Geometry
Chapter Forty Two – The Nature of Light
Chapter Forty Three – Light …Does It Travel Rectilinearly or Curvilinearly?
Chapter Forty Four – Distance To the Stars

Rational Science Vol IV

Chapter One – The Rational Scientific Method
Chapter Two – Rational Physics
Chapter Three – Experimenter's Regress
Chapter Four – Knowledge and Prediction
Chapter Five – Word Magic V1.1

Chapter Six – Nature of Scientific Inquiry
Chapter Seven – Karl Popper
Chapter Eight – Words Mean Things
Chapter Nine – Are Humans Intelligent?
Chapter Ten – Freewill - What Is It?
Chapter Eleven – Freewill – Part Two
Chapter Twelve – Higgs Boson – What Is It?
Chapter – Thirteen – The Higgs Fake
Chapter Fourteen – The Higgs Fake - Part Two
Chapter Fifteen – The Higgs Fake – Part Three
Chapter Sixteen – Radioactive Ion Beams
Chapter Seventeen – Tether Hypothesis
Chapter Eighteen – Sorce Theory and Other Fantasies
Chapter Nineteen – Relativity , Dirac's Equation , and The Color Of Gold
Chapter Twenty – Naked Ape
Chapter Twenty One – Relativity's Failed Predictions
Chapter Twenty Two – Expanding Earth Hypothesis
Chapter Twenty Three – Dinosaurs, How Could Some of Them Be So Big?
Chapter Twenty Four – Dinosaur Size Paradox
Chapter Twenty Five – Big Dinosaurs
Chapter Twenty Six – Dino Size Paradox Solved
Chapter Twenty Seven – CO_2 and Its Role In Global Warming
Chapter Twenty Eight – CO_2 and Its Role In Global Warming Part Two
Chapter Twenty Nine – CO_2 and Its Role In Global Warming Part Three
Chapter Thirty – CO_2 and Its Role In Global Warming Part Four
Chapter Thirty One – Rope Hypothesis
Chapter Thirty Two – Loop Theory? Hypothesis?
Chapter Thirty Three – Forces of Nature – Push and Pull
Chapter Thirty Four – Atom and Cell
Chapter Thirty Five – Why Fattie Gets A Sunburn
Chapter Thirty Six – Backscattering
Chapter Thirty Seven – Neutron Bombardment, Beta Decay,

and Radiation
Chapter Thirty Eight – Photoelectric Effect and Rope Hypothesis

Rational Science Vol. V

Chapter One – The Rational Scientific Method
Chapter Two – Intelligent and Smart
Chapter Three – Artificial Intelligence - Artilect
Chapter Four – Collaborative Consumerism
Chapter Five – Sapiens, a Brief History of Mankind
Chapter Six – Expanding Space Silliness
Chapter Seven – Infinitiverse
Chapter Eight – Where Is the Edge of the Universe?
Chapter Nine – Will Quantum Mechanics Swallow Relativity?
Chapter Ten – LIGO and Gravitational Waves
Chapter Eleven - Mach's Principle
Chapter Twelve – Plasma Tubes
Chapter Thirteen – Aether and IAAAD
Chapter Fourteen – Planetary Evolution
Chapter Fifteen – Stellar Metamorphosis
Chapter Sixteen – Transformation Hypothesis
Chapter Seventeen – Solar System Formation
Chapter Eighteen - Brown Dwarfs and Migrating Planets
Chapter Nineteen – Planetary Evolution and the Rational Scientific Method
Chapter Twenty – Ultra Cold Experiment
Chapter Twenty One – Black Body
Chapter Twenty Two – Black Body Radiation
Chapter Twenty Three – Rope Hypothesis
Chapter Twenty Four – Why Not Rope Theory?
Chapter Twenty Five – Atom and Cell
Chapter Twenty Six – Elementary and Composite

Chapter Twenty Seven – Which Came First?
Chapter Twenty Eight – Thread Theory
Chapter Twenty Nine – Black Body and Thread Theory
Chapter Thirty – What is a Shadow?
Chapter Thirty One – Light, Gravity, and Magnetic Moment
Chapter Thirty Two – Plasma the Fourth State of Matter?
Chapter Thirty Three – Ions, Charge and Matter
Chapter Thirty Four – Elementary Charge
Chapter Thirty Five – Batteries, Current Flow and Ions
Chapter Thirty Six – Atomic Bonding
Chapter Thirty Seven – Thousands of Kilometers Long Broomstick
Chapter Thirty Eight – How is Sound Different Than Light?
Chapter Thirty Nine – The World's First Flashdark
Chapter Forty – Gravitation and Electrostatics
Chapter Forty One – Antenna Theory and Rope Hypothesis

The Best of Rational Science

Chapter One – The Rational Scientific Method
Chapter Two – Scientist, Science, & The Scientific Method
Chapter Three – Scientific Method? For Dummies!
Chapter Four – Hypothesis, Theory, Conclusion
Chapter Five – Science & Technology - Conceptual & Empirical
Chapter Six – Experiments Are They Part of the Scientific Method?
Chapter Seven - Peer Review
Chapter Eight – Proof is For Alcohol
Chapter Nine – Rational Physics
Chapter Ten – Experimenter's Regress
Chapter Eleven – Knowledge and Prediction
Chapter Twelve – Word Magic V1.1

Chapter Thirteen – Nature of Scientific Inquiry
Chapter Fourteen – Karl Popper
Chapter Fifteen – Words Mean Things
Chapter Sixteen - Math
Chapter Seventeen– Life
Chapter Eighteen – The Sense of Touch
Chapter Nineteen – Temperature, What Is It?
Chapter Twenty – Dimensions
Chapter Twenty One – Dimensions of Reality
Chapter Twenty Two – The Three Dementia of Geometry
Chapter Twenty Three – Time Part One
Chapter Twenty Four – Time Part Two
Chapter Twenty Five – Time Part Three
Chapter Twenty Six – Mass Part One
Chapter Twenty Seven – Mass Part Two
Chapter Twenty Eight – Mass Part Three
Chapter Twenty Nine – Big Bang Theory
Chapter Thirty – Black Holes
Chapter Thirty One – The Nature of Light
Chapter Thirty Two – Light ...Does It Travel Rectilinearly or Curvilinearly?
Chapter Thirty Three – Distance to the Stars
Chapter Thirty Four – Shapiro Effect
Chapter Thirty Five – Distance ...The Rubber Ruler
Chapter Thirty Six – Relativity's Failed Predictions
Chapter Thirty Seven – What Happened to the Dinosaurs
Chapter Thirty Eight – Expanding Earth Hypothesis
Chapter Thirty Nine – Dinosaurs, How Could Some of Them Be So Big?
Chapter Forty – Dinosaur Size Paradox
Chapter Forty One – Big Dinosaurs
Chapter Forty Two – Dino Size Paradox Solved
Chapter Forty Three – Mach's Principle
Chapter Forty Four - Atom

Chapter Forty Five - Planetary Evolution
Chapter Forty Six – Stellar Metamorphosis
Chapter Forty Seven – Transformation Hypothesis
Chapter Forty Eight – Solar System Formation
Chapter Forty Nine – Brown Dwarfs and Migrating Planets
Chapter Fifty – Planetary Evolution and the Rational Scientific Method

Rope Hypothesis and Thread Theory

Introduction

Chapter One - Rope Hypothesis
Chapter Two - Rope Theory, Thread Theory?
Chapter Three - Loop Theory? Hypothesis?
Chapter Four - Fundamental and Composite
Chapter Five -
Chapter Six - Cyclical Time or Circular Reasoning?
Chapter Seven - Causality and Inevitable Recurrence
Chapter Eight - Self Organizing Systems, Thermodynamics, Gravity and Life
Chapter Nine - General Systems Theory
Chapter Ten - Thermodynamics and Living Systems
Chapter Eleven - Thermodynamics and Evolution on a Cellular Level
Chapter Twelve - From Molecules to Galaxies
Chapter Thirteen - Thermodynamics on a Cosmic Scale
Chapter Fourteen - The Laws of Eternal Dynamics
Chapter Fifteen - Atomic, Nano, Micro, Macro and Cosmic
Chapter Sixteen – Friction
Chapter Seventeen - Types of Atomic Motion
Chapter Eighteen - Which Came First?
Chapter Nineteen - Bending and Stretching Thread
Chapter Twenty – Tension
Chapter Twenty One - What is a Shadow?

Chapter Twenty Two - Thousand of Kilometers Long Broomstick
Chapter Twenty Three – Sine Waves
Chapter Twenty Four - How is Sound Different than Light?
Chapter Twenty Five - Light and Sound, How Are they Different?
Chapter Twenty Six - The World's First Flashdark!
Chapter Twenty Seven - Light, Gravity and Magnetic Moment
Chapter Twenty Eight - Batteries, Current Flow and Ions
Chapter Twenty Nine - Cathode Ray Tube
Chapter Thirty - Closed Circuit
Chapter Thirty One - Plasma, the Fourth State of Matter?
Chapter Thirty Two - Ions, Charge and Plasma
Chapter Thirty Three - Elementary Charge
Chapter Thirty Four - Gravitation and Electrostatics
Chapter Thirty Five - Antenna Theory and Rope Hypothesis
Chapter Thirty Six - Gravity Basics
Chapter Thirty Seven - Big G in Newton's Law of Gravitation
Chapter Thirty Eight - Big G Two
Chapter Thirty Nine - Tension as Numbers
Chapter Forty - Pumping Atoms
Chapter Forty One - Light Propagation
Chapter Forty Two - Refraction, Reflection and Diffraction
Chapter Forty Three - Superimposition of Ropes and Waveforms
Chapter Forty Four - Spinning Atom
Chapter Forty Five - Blackbody Radiation and Thread Theory
Chapter Forty Six - Atomic Bonding

www.ingramcontent.com/pod-product-compliance
Lightning Source LLC
Chambersburg PA
CBHW071528220526
45469CB00003B/690